FLOATING GATE DEVICES: OPERATION AND COMPACT MODELING

Floating Gate Devices: Operation and Compact Modeling

by

Paolo Pavan
DII – Università di Modena e Reggio Emilia,
Italy

Luca Larcher
DISMI – Università di Modena e Reggio Emilia,
Italy

and

Andrea Marmiroli
STMicroelectronics,
Central R&D, Italy

KLUWER ACADEMIC PUBLISHERS
BOSTON / DORDRECHT / LONDON

A C.I.P. Catalogue record for this book is available from the Library of Congress.

ISBN 978-1-4419-5426-8   e-ISBN 978-1-4020-2613-3

Published by Kluwer Academic Publishers,
P.O. Box 17, 3300 AA Dordrecht, The Netherlands.

Sold and distributed in North, Central and South America
by Kluwer Academic Publishers,
101 Philip Drive, Norwell, MA 02061, U.S.A.

In all other countries, sold and distributed
by Kluwer Academic Publishers,
P.O. Box 322, 3300 AH Dordrecht, The Netherlands.

Printed on acid-free paper

Contents

3. DC CONDITIONS: READ   37

4. TRANSIENT CONDITIONS: PROGRAM AND ERASE   57

Contributing Authors

Paolo Pavan graduated in Electrical Engineering at the University of Padova, Italy, in 1990 working on latch-up and hot-electron degradation phenomena in MOS devices. In 1991 he started his PhD program studying impact ionization phenomena in advanced bipolar transistor and received his PhD in 1994. From 1992 to 1994 he has been at the University of California at Berkeley where he studied radiation effects on MOS devices and circuits. Recently his interest moved to the characterization and modeling of Flash memory cells and on the development of new nonvolatile cells. He worked with Saifun Semiconductors (Israel) on the optimization of NROM memory devices. He authored and co-authored technical papers, invited papers and two chapters in books.

He is currently Associate Professor of Electronics at the University of Modena and Reggio Emilia, Italy.

Luca Larcher graduated in Electronic Engineering at the University of Padova, Italy, in 1998 working on the modeling of gate oxide currents in MOS devices. In 1998 he started his PhD program working on the compact modeling of non-volatile (EEPROM and Flash) memories. He received his PhD in January 2002. His research interests concern the reliability, the characterization and the compact modeling of MOS and Non-Volatile memories (Flash, EEPROM, NROM). In this field, he focused on the modeling of stress and radiation induced leakage currents, Channel-Hot Electron currents, and MOS gate capacitance. He authored and co-authored technical papers.

He is currently Researcher of Electronics at the University of Modena and Reggio Emilia, Italy.

Andrea Marmiroli graduated in Physics at the University of Milano, Italy, in 1984 with a thesis work on an algebraic approach to quantum field theory in theoretical physics. In 1985 he joined STMicroelectronics, where he worked at first in the New Technology Development Group of the Central R&D on Microlithography. In 1989 he joined the Technology CAD group. During this activity he was author of a few papers in the process simulation area. He has been also responsible for the development of the architecture of the MOS transistors architecture for the 0.8 micron Flash EEPROM memory process. He has been involved as technical responsible for STMicroelectronics in many European funded projects in the TCAD area. He coordinated 3 thesis works in physics and electrical engineering and one PhD Thesis. Since 1995, besides leading the Technology CAD activity in Agrate, he is responsible of modeling and parameter extraction for Spice-like simulation.

Preface

The goal of this book is twofold. First, it explains the principles and physical mechanisms of Floating Gate device operations. Second, starting from a general overview on Compact Modeling issues, it illustrates features and details of a complete Compact Model of a Floating Gate device, the building block of Flash Memories, one of the "hottest" products in the semiconductor industry.

Flash Memories are one of the most innovative and complex types of high-tech, nonvolatile memories in use today [see, for example, Proceedings of the IEEE, Special Issue on: *Flash Memory Technology*, April 2003]. Since their introduction in the early 1990s, these products have experienced a continuous evolution from the simple first products to emulate EPROM memories, to the extreme flexibility of design application in today products. This is an enabling technology: future limits are beyond our current expectations and limited only by our imagination.

In the memory arena, Flash memory is the demonstration of the pervasive use of new electronic applications in our lives. Every new application can exploit this flexible and powerful memory technology, either as a stand-alone component or integrated as the enabling feature of the whole silicon integration.

Flash are not just memories, they are "complex systems on silicon": they are challenging to design, because a wide range of knowledge in electronics is required (both digital and analog), and they are difficult to manufacture. Physics, chemistry, and other fields must be integrated; and conditions must be carefully monitored and controlled in the manufacturing process.

Memories demand massive investments in R&D, but they also reward with enormous potential market values. Flash memory market (considered the most important market segment among nonvolatile memories) is expected to progress at a very fast pace, and to gain the second place in the overall memory market. This is due to the optimization of cost/performance tradeoffs, and in particular to the inherent flexibility and versatility of this memory, which brings benefits in many applications.

The leading application is in multimedia systems, which require memories that are increasingly larger in size, and demand ever-increasing performance characteristics.

Telecommunications, computers, automotive and consumer electronics are some additional areas where these memories make possible numerous emerging applications.

Moreover, the Flash memory integration is one of the irreplaceable requirements for further technological innovations, and particularly to realize the so-called *system on silicon*.

Compact Model (CM) means an analytic model of the electrical behavior of a circuit element. Modeling is usually aimed at providing means to simulate the behavior of a device or a circuit by quantitative calculation. CM allows to highlight basic properties of a device, thus making easier the understanding and the synthesis of robust circuits. Therefore, the main intent of modeling is to forecast the behavior of a system. This holds for all integrated devices (resistors, capacitors, inductors, transistors, and also the device subject of this book: the *floating gate device*) and circuits.

Compact Models of Floating Gate devices have the same purpose of all compact models: *to be used within a program for circuit simulation*. The Floating Gate transistor is the building block of a full array of memory cells and a memory chip. In a first approximation, the reading operation of a FG device, and for some cases also programming and erasing, can be considered a single-cell operation. Nevertheless, CMs are fundamental to simulate the effects of the cells not directly involved in the operation under investigation and the effects of the parasitic elements. Furthermore, they allow the simulation of the interaction with the rest of the device, and hence they are useful to check the design of the circuitry around the memory array: algorithms for cell addressing, charge pump sizing taking into account current consumption and voltage drops, etc...

In addition, CMs are expected to become more and more important in the forecasted scenario of semiconductor industry, where few major manufacturing foundries with large capacity will produce wafers for many different design centers, each one designing their own products based on diverse simulation tools. In this picture, CMs play a central role: they link manufacturers to designers, and they are vital for a correct implementation of the design in silicon avoiding as much as possible any return to production line due to poor matching between the final products on silicon and simulation predictions. In this context, having CMs capable to adapt easily to different technologies by means of a restricted number of parameters (possibly easy to extract) will surely become a great advantage.

Finally, CMs are essential to progress to an easier and faster development process of new Non-Volatile Memory products. CMs are the bridge between process and design: they simulate the device behavior, which depend on how devices are manufactured (that is, the process recipe), in a fashion that it is easy to understand for designers, who use CMs to design and calibrate circuits. For this reason, it is reasonable to forecast that CMs will play a progressively more important role in the future semiconductor scenario.

In this scenario, despite of the wide diffusion of FG-based Non-Volatile Memories, no complete CMs of FG devices were proposed and used in the industry until few years ago. Usually, MOSFET transistor whose threshold voltage were *manually* changed to model programmed and erased state of the FG memory cell, were used in circuit simulations to reproduce (with poor accuracy) the FG memory behavior.

The motivation for our work in the last years has been just to fill this gap. Now, a compact model capable to simulate read/program/erase operations of FG devices is available, and an implementation of it into a commercial circuit simulator is currently used by R&D people in STMicroelectronics.

In this book, the approach followed and specific details of the developed CM are widely described, giving also a general overview on FG device operation and CM issues. A list of chapters follows, along with an explanation of their content and of their purpose.

Chapter 1: an introduction on Compact Modeling and Semiconductor Memories will be given, to create a common background.

Chapter 2: the principles of Floating Gate devices will be given, starting from technology highlights, to cell operations, physical aspects and reliability issues.

Chapter 3: after an overview on what proposed in the literature, a new CM approach is proposed. Then, the compact model of a FG device in DC conditions developed for read operation simulations will be illustrated.

Chapter 4: program and erase operations will be analyzed, describing their physical mechanism and explaining in details issues and solutions for an effective compact modeling.

Chapter 5: further possibilities of this new CM will be proven. Reliability predictions and statistical simulations will be introduced.

Chapter 6: in this chapter, Flash memories will be described from a designer point of view. The whole product will be analyzed, and the role of CM to design a challenging memory product will be highlighted.

Paolo Pavan
Luca Larcher
Andrea Marmiroli

Foreword

This book gives our personal vision on Compact Modeling of Floating Gate Devices. It collects the results of our research activity in these last years, some of which have already been published, after peer review process, in international journals.

The philosophy we have adopted to develop the Compact Model of Floating Gate devices proposed in this book is based on a modular approach. The Floating Gate device is depicted as a simple subcircuit where device operations and functionality are simulated by specific current generators, each one describing through compact formula the current induced by the single physical phenomenon. This modular approach allows different levels of accuracy and complexity of the model, and leaves the door open for any improvement and change to tailor the model to new and different technologies.

In this book, the specific example of the Compact Model of a Floating Gate device is illustrated. Besides this specific content dedicated to Non-Volatile Memory people, we believe that the modular philosophy followed here could be a successful strategy for compact modeling, exported also to other devices. We hope that our personal vision of Compact Modeling can be shared by many in different areas of Semiconductor Industry, where compact modeling is important.

Paolo Pavan
Luca Larcher
Andrea Marmiroli

Chapter 1

INTRODUCTION

Compact Modeling and Simulation in Semiconductor Industry.
Semiconductor Memories.

In this book, the discussion on Compact Modeling (CM) of semiconductor memories based on Floating Gate (FG) device will be based on the understanding of few concepts, that will be explained in the next chapters.

In this introductory chapter, we will introduce basic concepts of CM and Semiconductor Memories based on the FG devices.

General notions and requirements of CM will be discussed along with the specific issues, advantages and limitations that must to face to develop a compact model of Floating Gate device.

We will give an overview on semiconductor memories and specifically on nonvolatile memories. Starting from the origin of the FG concept and the first working FG device (FAMOS), we will go on with the first commercial products and few of the evolutions that have survived these years. The material presented here is a general overview and it has the purpose of creating a common understanding before entering the details of operations and compact modeling of FG devices.

1. COMPACT MODELING

In this introductory paragraph, we will explain what Compact Modeling means, by illustrating and explaining some of the general concepts and definitions. We will discuss some of the requirements to face when developing a new model, specifically dealing with complexity, hierarchy and standardization.

The specific issues that need to be considered when developing the compact model of the device subject of this book, i.e. the *floating gate device* will be introduced, discussing implementation, advantages and limitations. Some remarks on possible future developments in the CM research area will be given.

1.1   General concepts and definitions

Modeling and characterization are often used together, being closely linked, nevertheless they have different meanings. In the semiconductor industry and in particular in the area of circuit design, to focus on the subject of this book, *Compact Model* (CM) means a model of the electrical behavior of a circuit element that does not require excessive computation effort, while *characterization* means the procedure by which the parameters of a compact model are determined for devices in a given Integrated Circuit (IC) manufacturing technology.

Modeling is usually aimed at providing means to simulate the behavior of a device or a circuit by quantitative calculation. Another important purpose consists in highlighting basic properties of a device, in order to facilitate the understanding and the synthesis of robust circuits [1]. Therefore, the main intent of modeling is to forecast the behavior of a system. This holds for all integrated devices (resistors, capacitors, inductors, transistors, the device subject of this book: the *floating gate device*) and circuits.

Back in the 60's, "ancient times" in electronics, to keep the pace of the growing complexity of electronic circuits it was necessary to give a tremendous improvement to the simulation techniques. The solution came from electronic computers: it was during those years that many simulation programs were developed, allowing overcoming the otherwise unaffordable complexity [2]. SPICE (Simulation Program with Integrated Circuit Emphasis) was the main result of this development activity, carried out at the University of California at Berkeley. Nowadays SPICE-like simulation programs are basic tools for circuit design: it cannot even be expected to design today's circuits without such tools. We can say that SPICE is now *the language* of circuit designers.

The implementation in a computer program is carried out to exploit the advantage of computers with respect to human mind: the capability to perform much faster calculations. All the know-how, all the smartness is in the model and in its implementation: the computer is just the tool to perform fast calculations.

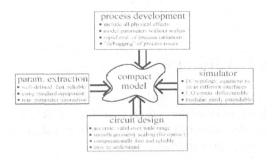

Figure 1-1. Compact model formulation [3].

The reasoning behind the development of a new CM is complex; its formulation is influenced by often contradicting requirements from basically four sides, as shown in Fig. 1-1: circuit design (the end user), device physics due to technology advances, parameter extraction, and simulator implementation. As a consequence, the resulting overall model form can only be a compromise, depending on the goals and priorities set during development, and most likely it will serve different applications differently well. A single model form is very unlikely able to meet different practical requirements [3].

Therefore, to reach the goal of reliably forecasting the behavior of a device and of a full circuit, CMs have to show many features and satisfy many constraints.

First, they should be physically formulated as function of both the fundamental process parameters that control device electrical behavior, and geometric layout parameters associated with a device (both adjustable layout parameters, such as device length and width, and technology dependent layout parameters, derived from design rules, such as spacing between active areas and implant areas, etc.). In facts, in the very early design phase of a new technology, test structures on silicon might not be available and hence model parameters have to be considered "target", i.e. not based on measurements. The generation of such models is possible only via TCAD simulations or modifying physically meaningful parameters.

Second, CM design involves a careful tradeoff between accuracy and efficiency. The model should be based on analog simulation requirements, which are usually more demanding than digital simulation requirements. Enough detail must be captured for circuit behavior to be well approximated, both quantitatively and qualitatively. Minimizing error is important, but the model should also be transparent enough that designers trust the results [4].

Looking at the MOSFET case, here the essence of CM is given also by a tradeoff among continuity, symmetry, scalability, predictability. The ability of a compact model in predicting electrical characteristics with accuracy in the second-/third-order derivatives from long/wide-channel down to threshold voltage roll-off regime is a major challenge, which is important for capturing geometry variations due to process fluctuations, for predicting new technologies, as well as for analog circuit design [5].

Third, the model implementation must be efficient and compact enough to allow large circuit to be simulated. The model equations should go as far as possible with clean physics, and then they have to use well-behaved empirical relations. Including as much physics as appropriate is likely to improve the ease of extraction, reduce the number of fitting parameters, and make the model better at extrapolation or scaling. Simplifying assumptions should be tested against data and/or detailed numerical simulations. The model needs to be fast and robust. Depending on the simulator in which the model is used, overshoot avoidance or other implementation features may be needed to guarantee proper convergence in the simulator [4].

When developing a new model, it is mandatory to keep in mind that CMs are complex. Complexity is defined by the number of nodes and elements in the equivalent circuit, along with the element equations and number of parameters. Moving towards complicated models gives probably better results, but leads to less and less understanding within the design community. Thus, model complexity should be lower or higher depending on the application of the model. Therefore, providing a model hierarchy is attractive.

In principle, a model hierarchy could be realized by choosing a specific reference model form and using flags to turn on and off certain effects. A medium complexity model should be used as reference, being the most sophisticated too heavy for some kind of simulations and the elements of its equivalent circuit too interrelated to allow any simplification [3].

There are obstacles to adopt a CM as standard in the design community. Assuming a working set of model equations in a SPICE-like simulator, the major obstacles are the implementation into commercial simulators, and the model parameter extraction. Both issues are of little interest to most of circuit designers, but they do impact model performance significantly. Unfortunately, parameter extraction is not a push-button task. A successful parameter determination requires: (i) understanding of the model and of its limitations: (ii) the development of an adequate methodology and associated extraction methods; (iii) the development and design of proper test structures [3].

Moreover, the current practice is to use the industry standard model, BSIM [6] on the golden die of a given technology by fitting a large number of parameters on a large set of measured electrical data. The model is optimized at the desired feature geometry, and not meant to be extrapolated outside the fitted range. But, if a CM fitting parameters (no matter how many) have to be obtained through global optimization without physical relation to process variation, CM may fit one transistor perfectly, but it may lose predictability for other transistors in the given technology [5].

1.2   The Compact Modeling of a Floating Gate device

Many models for elementary devices have been developed during these last 40 years, some with emphasis on physics, some on behavior, all keeping into account that the model had to be implemented in a computer program to forecast the behavior of a full circuit.

Compact models of floating gate devices have the same objective: *to be used within a program for circuit simulation*. The FG device CM is the building block to model a full array of memory cells and a memory chip. In a first approximation, the reading operation, and for some cases also writing and erasing, can be considered a single-cell operation. Nevertheless, a compact model allows the simulation of the effects of the cells not directly involved in the operation under investigation and the effects of the parasitic elements. Moreover, it allows the simulation of the interaction with the rest of the device, and hence it is useful to check the design of the circuitry around the memory array: algorithms for cell addressing, charge pump sizing taking into account current consumption and voltage drops, etc...

The starting point of the FG device model presented here [7] is the MOS (Metal Oxide Semiconductor) transistor model. Many MOS models have been developed and many are currently available (Philips MM11 [8], BSIM4 [9], EKV [10], SP [11], HiSIM [12]). The approach presented here is independent on a specific choice. Therefore, one of the main advantages of this approach is the exploitation of all the improvements carried out for the basic MOS transistor and, as it will be shown later, the definition of the relative parameter extraction algorithms.

The simple idea underneath is to model the floating gate device as a circuit with a MOS transistor and a capacitor between the control gate and the floating gate node, which is the gate of the MOS transistor, Fig 1-2 [7]. Actually the capacitor is not included in the implemented model, but its capacitance value is used, together with the charge injected in the floating gate, to calculate the FG node potential that is applied through a voltage controlled voltage source (V_{FG} in Fig. 1-2). There is no general solution to the calculation of the potential of a floating gate node in a DC circuit simulation. This is the reason to add the voltage source device to force the potential of the floating gate node in DC simulations.

This model has also the advantage to allow the modeling of programming and erasing operations by simply adding a set of suitable current generators between the various electrodes. This modular approach permits the modeling of read disturbs, retention, leakage currents, in a rather simple way. Besides, it allows the development of different models for all these current generators (for program, erase, charge loss,...) that can be compared one with the other, and that can be enhanced or changed to take into account new mechanisms or technologies.

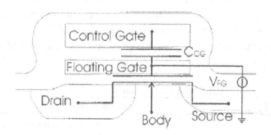

Figure 1-2. Schematic of the new compact model of a FG device [7].

Two main limitations of this model can be foreseen. First, usually MOS compact models target thin gate oxide transistors with Lightly– or Medium– Doped Drain (LDD / MDD) diffusions. The oxide thickness of all the current floating gate devices is above the 7 nm, while the source and drain junctions are usually abrupt. It might become necessary to adapt the existing transistors models to this kind of devices. Second, there is a bounce of coupling capacitances which is neglected: the coupling between the control gate node and the source, drain, and body nodes.

Furthermore, as the memory cells are getting smaller and closer one to the other, the coupling capacitance between the electrodes of two neighbor cells (which are not included in the model) are getting more important. We think that there will be the need to take them into account and hence the need of a smart solution to the problem of modeling this capacitive net.

To conclude, it is worth mentioning that the forecasted scenario of semiconductor industry predicts the existence of few major manufacturing foundries with large capacity that produce wafers for many different design centers, each one designing their own products based on several simulation tools. In this picture, CM will play progressively a more central role: it links manufacturers to designers, and it is vital to a correct implementation of the design in silicon avoiding as much as possible any return to production line due to poor matching of results on silicon and simulation predictions.

2.   SEMICONDUCTOR MEMORIES

Artificial memories share with human memory the following objectives: fast storage of information, fast retrieval of the information stored; possibility of recording a large amount of data; never forget, unless wanted. For artificial memories, depending on the application, one characteristic will be more important than the others, but one aspect will always be dominant: cost should be the lowest possible.

The semiconductor memory market is not an exception: these characteristics have always been pursued and clearly some trade offs are necessary, as there is no perfect solution. Semiconductor memories have been around now for many decades. Their density has continuously increased (almost four times every three years) and they are used in many new applications, where both high speed and low power are demanded.

Complementary metal-oxide-semiconductor (CMOS) memories can be divided into two main categories: random access memories, RAMs (which are volatile, i.e., they lose stored information once the power supply is switched off), and read-only memories, ROMs (which are nonvolatile, i.e., they keep stored information also when the power supply is switched off).

Table 1-1. Memory types and related main features.

Memory Type	Features
FLASH	Low-cost, high-density, high-speed architecture; low power; high reliability
ROM Read-Only Memory	Mature, high-density, reliable, low cost; time-consuming mask required, suitable for high production with stable code
SRAM Static Random-Access Memory	Highest speed, high-power, low-density memory; limited density drives up cost
EPROM Electrically Programmable Read-Only Memory	High-density memory; must be exposed to ultraviolet light for erasure
EEPROM or E²PROM Electrically Erasable Programmable Read-Only Memory	Electrically byte-erasable; lower reliability, higher cost, lowest density
DRAM Dynamic Random Access Memory	High-density, low-cost, high-speed, high-power

"While the price has dropped from one dollar per 100 bits to less than one dollar per 100 megabits, the production technology has become so complex and expensive that the average multibillion dollar company can no longer afford it. Circuits have reached sub-micrometric geometries, and also reliability has been impacted as the reduced storage capacitance together with charge loss phenomena require to design memory systems with error correction circuits. The applications have migrated from a few, large, expensive industrial computers to a wide range of low cost, ubiquitous, consumer handheld devices.

The result is that the few, high volume, cost effective, commodity memory products of the past have fragmented into a range of application specific devices with increasing production and design costs each with its own mix of speed, power, random access and data retention. Meanwhile, the dream of a true non-volatile, fast, low power, random access memory has never been achieved." [13]

The ideal memory subsystem optimizes density, preserves critical material in a nonvolatile condition, is easy to program and reprogram, can be read fast, and is cost-effective for the application. Some memory technologies meet one or more of these requirements very well, but offsetting limitations can prevent the product from becoming a genuine solution, especially in newer applications.

The application specific devices result in a widespread variety of Non Volatile Memories (NVMs), and they all show different characteristics according to the structure of the selected cell and the complexity of the array organization. They all have performance that can go from those of ROM memories, which cannot be reconfigured, to those of information alterability with almost the same flexibility of RAM memories, see Table 1-1.

3. FLOATING GATE DEVICES

FG devices are at the core of every modern NVMs. The basic concepts and the functionality of this kind of device are easily understood if it is possible to determine the FG potential. The schematic cross section of a generic FG device is shown in Fig. 1-3: the upper gate is the control gate and the lower gate, a conductive layer completely isolated within the gate dielectric, is the FG. The FG acts as a potential well (see Fig. 1-4). If a charge is forced into the well, it cannot move from there without applying an external force: the FG stores charge.

More than 90% of non volatile memory production is based on the Floating Gate concept [14]. Despite of the many improvements since its first introduction many years ago [15] -the third generation of Flash Memories, enabled by improved processes, new writing mechanisms and new architectural solutions is behind the corner-, Flash Memories (and therefore the FG device) continue to play the most important role in the NVM arena.

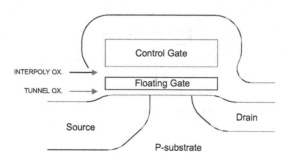

Figure 1-3. Cross Section of a Floating Gate Device.

To have a memory cell that can commute from one state to the other (from "programmed, 0" to "erased, 1") and that can store the information independently of external conditions, the storing element needs to be a device whose conductivity can be changed in a nondestructive way.

The FG memory device is an MOS transistor with a threshold voltage that is given by:

$$V_T = K - Q_{FG}/C_{CG} \qquad (1)$$

where K is a constant that depends on the gate and substrate material, doping, and gate oxide thickness, Q_{FG} is the charge in the Floating Gate, and C_{CG} is the capacitance between Control Gate and Floating Gate [16].

The threshold voltage can be altered by changing the amount of charge present in the floating gate. Thus, by storing/removing charge in/from the floating gate, the threshold voltage can be changed repetitively from a high to a low state (following the P1005 IEEE Draft Standard for Definitions, Symbols and Characterization of Floating Gate Memory Arrays [17], "programmed" and "erased" states, respectively).

The "read" operation is performed by applying a gate voltage that is between the values of the erased and programmed threshold voltages and sensing the current flowing through the device. In Fig. 1-5, transcharacteristic of a FG device in "programmed" and "erased" states are shown.

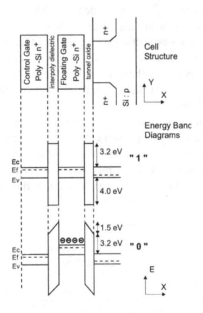

Figure 1-4. Cross section and band diagram of a FG Device.

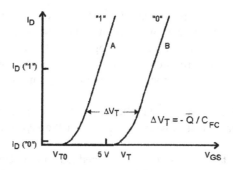

Figure 1-5. Current-Voltage transcharacteristic of a FG device when there are no electrons stored (curve A, "1") and when charge is stored (curve B, "0") in the FG.

4. FIRST COMMERCIAL DEVICES AND PRODUCTS

The memory cell, considered as a single device, will operate in a memory bench, which is composed of an array of cells, in its simplest case. In this array, cells which are on the same column share the same source/drain diffusion (Bit Line, BL), and cells on the same row are connected through the same polycrystalline silicon line which forms the gates of devices (Word Line, WL). The information stored in each single device can be accessed (read or written) by applying the correct voltages to the BLs and WLs of the memory array.

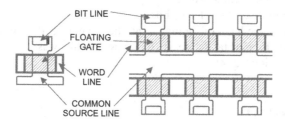

Figure 1-6. FG cell layout and NOR Array Architecture.

It is important to stress how the cell, considered as a single device, differs significantly from the cell considered as an element of the array. Access to the single array element is now performed through other elements of the circuitry, either on the same column or on the same row. Moreover, the operations to be performed on a single element of the array have an influence on the elements of the same row/column.

The first floating gate devices where produced in the 70's. It is surprising to think how different from now the technology was in those days.

First of all the dimensions of the wafers: 3 inches, compared to the current 300 mm (12 inches); but also the minimum dimensions of transistors' lengths, that was around 7 μm, i.e. about 70 times today's minimum dimensions; and the maximum program and erase voltages that reached the 30 Volts.

Source and drain junction depth was 1.5-2.0 µm, while the gate oxide of the transistors and of the memory cell was around 100 nm: there was already a full compatibility between the memory part and the logic part. The dimension of the memory cell was a few hundreds of square micron, and included the floating gate transistors and the select transistor. For these first processes, the number of masks ranged from 8 to 11 masks.

Compared to the current layout techniques, including all the software tools for the automatic synthesis till layout, it is worth noticing that the layout in the 70's was hand drawn on paper.

As an example let's take a system with embedded non volatile memories: a TV synthesizer developed and manufactured by STMicroelectronics -SGS, in those times- [18-21]. The function of the device was the storing of all the information necessary to memorize a set of channels. The first devices were produced in 1978 with a die size of around 5 mm × 4 mm, a cell size of 80 µm × 40 µm, the total memory capacity was 272 bits. The total number of components was about 5000. The voltage necessary to program and erase was 29 V, with nodes that could reach 45 V. Cells were programmed by hot carriers while the erasing was by Fowler-Nordheim between the two polycrystalline silicon layers (through the "interpoly" oxide). It is interesting to underline some characteristics that are currently routinely applied: the erase-verify (the erase operation was performed on the whole word as part of the word programming) and the program verify for the single bit, so that the programming of each single bit was stopped as soon as the target threshold was reached. Actually the programming flow was:

1. program of all the bits, with the mentioned bit-by-bit verify;
2. erase of all the whole word;
3. programming of the selected bits.

This allowed the same level of "aging" of all the memory cells within the same word. This device was able to withstand at least 10.000 cycles: this was the required number of cycles for the application [22-24].

5.   EVOLUTION

Getting back to today's standard NOR architecture, where the access to the stored data is addressed by biasing two of the three terminals of the single cell (being the third a common reference to the whole array or part of it), a way to increase the capacity is also achieved following other methodologies to store and read data.

New architectures, for example NAND, have been introduced. They are based on the same single storage element, which is connected and accessed in different ways to achieve larger capacity per unit area. The elementary unit is not composed of the single three-terminal cell, which stores one single bit, but by more FG transistors connected in a series (8 or 16) which constitutes a chain connected to the bit line and ground through two selection transistors, see Fig. 1-7. This organization allows the elimination of all contacts between WLs, thus reducing the occupied area by 40%, with respect to NOR architectures. Moreover, a kind of memory organization with a unit element with the dimension of one byte (or one word) is closer to the ideal memory with parallel access. It allows even page (256-byte) programming, resulting in a greatly improved versatility.

This new architecture is used for mass storage applications, where a large storage capacity is required, but slower read times and maybe some larger storage errors can be tolerated.

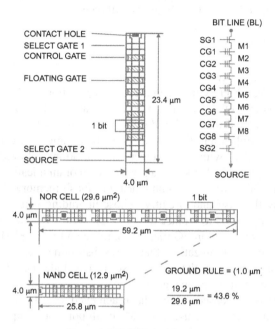

Figure 1-7. NAND Architecture.

Another approach is to store more bits in the same single FG device. Some manufacturers have two Flash technologies based on the same core memory cell. The first technology is the original single-bit-per-cell Flash memory that allows a single bit of information to be stored in each cell. The second and more recent technology is called multi-level cell structure (e.g., used in Intel StrataFlash® memory). This technology allows two bits of information to be stored in a single FG transistor, see Fig. 1-8. Precision is the key to storing reliably two bits per cell. Writing a cell (charge placement/removing) and reading (sensing) must be precisely controlled in order to have four states within a single FG transistor.

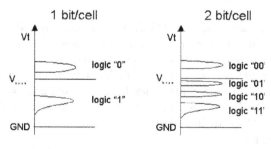

Figure 1-8. Single Bit and Multilevel storage in a single FG device

6.   APPLICATIONS AND MARKET CONSIDERATIONS

A very short summary of current applications of Flash memories and an overview of the 2002 sales data and of 2003-2010 forecasts will be given.

6.1   Applications

Even if the border between the areas covered by the different memories is continuously changing, currently the main areas of applications of NVMs are the storage of: (i) Operating System (OS) and (ii) any kind of data used by any kind of program. These two applications are usually referred to as "code" and "data".

As for the "code" application, the advantage of being able to electrically change the programs to be executed with low power consumption, allowed Flash-EEPROM memories to occupy the whole "code" area for wireless communications. In this case, e.g. in the chip for Bluetooth™ wireless communication, Flash memories store the firmware of protocols used by the baseband circuitry, whose upgrading versions are continuously released and need to be stored to replace the older protocols. Besides this use, in cellular phones also data are stored within FG memories, more and more Flash memories with a dedicated memory array organization that are substituting the EEPROM memories usually used for this purpose in the past. These data include the address book, simple calendar managers, setup of the phone, games, etc.

New generations of FG-based devices have achieved very low power consumption: this is also the reason for the large use of NVMs in all portable equipments both for code and data.

We want to emphasize here that data can be very different depending on their application, thus demanding for different memory array organization. For example: 1) images and MP3 files will be not changed, and likely will be erased as a whole, so they suite Flash memories with block erase capability; 2) the address book of one's PDA or the car's driver seat setup requires a memory with byte erase capability, and therefore an EEPROM with will be the best choice.

One additional note: historically RAM has been used to store the status of the execution of a program. This could change, as a NVM could be used in this case and allow program execution restart after a switch off from the exact switch-off situation.

Today the main market of FG memories is portable equipments, and the applications span from set top boxes, cars, games, computers, peripherals, etc. in all cases mainly for "code".

One of the greatest advantages of FG devices is that it is CMOS compatible. This allows the processing of a single chip containing both a processor and a memory array. In this case we call these memory arrays *embedded memories*.

The following considerations are common to the different kind of memories, not only the floating gate ones. Embedded memories have pros and cons with respect to other solutions, like "two chips" or "system in package." The main advantages are:
- Better Electro-Magnetic Immunity
- Less interface logic
- Faster systems
- Lower consumption

- Lower manufacturing costs

While the disadvantages are:
- Design times and costs are 30 to 50% higher
- More complex process
- Higher test costs
- Lower yield

The main applications for embedded non volatile memories are in the automotive segment, but with a high potential in the field of portable electronics.

6.2 Market highlights

Let's now give some figures about NVM market. In the last years (2000-2002) the total semiconductor market has been in the range 150-200 b$, with a NVM market in the range of 10-15 b$, i.e. 7% of the market.

These –approximated– values are divided according to applications and memory family in Table 1-2.

Table 1-1. NVM market in 2001 and expectations for 2003.

		2001 rev. M$ (appr.)	2003 rev. M$ (exp.)
EPROM		500	400
EEPROM		700	1000
Mask ROM		700	400
FLASH EEPROM	*NOR*	*5000*	*6000*
FLASH EEPROM	*Nand*	*1000*	*3500*
FLASH EEPROM	*Combo*	*2000*	*2500*
FLASH EEPROM	Total	8000	12000

The main application area for Flash EEPROM is, and will be also for the coming years, communication, as we can see in the following Table 1-3.

Table 1-2. Major applications of Flash EEPROMs.

	2001 rev. M$ (appr.)	2003 rev. M$ (exp.)
Communication	5100	6600
Industrial	600	700
PC & peripherals	1000	1500
Transportation	550	1200

From the two tables above it is clear that the most important use of floating gate device is in Flash EEPROM memories. This is why in this book we are going to focus on many aspects of CM of this kind of device. Among the different Flash architectures, currently the most important is the NOR one. The market best seller has been the 16Mbit in 2001 and the 32 Mbit in 2002 that will continue also in 2003.

The second important architecture is the NAND whose best seller is now the 256Mbit. The NAND architecture is dedicated mainly to data storage and it is expected to oversell

NOR architecture in the next coming years. NAND acceleration will be driven by *Flash cards* with a plethora of possible applications: digital cameras, MP3, data storage for PDA …

The average selling price (ASP) for Flash EEPROM has been around 5 $ for years. The growth in the memory capacity implies a reduction of the cost per bit.

To conclude, in the next Table 1-4 we want to show the share among the production costs of low and high density, stand alone and embedded floating gate devices

Table 1-3. Production costs for different kind of FG-based NVMs.

	Low dens. Standalone	High dens. Standalone	Low dens. Embedded	High dens. Embedded
Production Fab	65%	45%	60%	40%
Wafer testing	15%	20%	17%	22%
Package	5%	10%	5%	10%
Final test	15%	25%	18%	28%

REFERENCES

1. E. Vittoz, C. Enz, F. Krummenacher, "A basic property of MOS transistor and its circuit implementation" Proc. of Workshop on Compact Modeling, in Nanotech 2003, Feb. 23-27, S. Francisco, CA, USA, Vol. 2, www.nsti.org, ISBN 0-9728422-1-7, pp. 246-249.
2. Daniel Foty, "MOSFET MODELING WITH SPICE," Prentice Hall,1997 ISBN: 0-13-227935-5.
3. M. Schrroter, "Compact bipolar transistor modeling: issues and possible solutions", Proc. of Workshop on Compact Modeling, in Nanotech 2003, Feb. 23-27, S. Francisco, CA, USA, Vol. 2, www.nsti.org, ISBN 0-9728422-1-7, pp. 282-285
4. D.A: Bell, K. Singhal, H.K. Gummel, "USIM design considerations," Proc. of Workshop on Compact Modeling, in Nanotech 2003, Feb. 23-27, S. Francisco, CA, USA, Vol. 2, www.nsti.org, ISBN 0-9728422-1-7, pp. 250-253.
5. X. Zhou, S.B. Chiah, K.Y. Lim, "A technology-based compact model for predictive deep-submicron MOSFET modeling and characterization", Proc. of Workshop on Compact Modeling, in Nanotech 2003, Feb. 23-27, S. Francisco, CA, USA, Vol. 2, www.nsti.org, ISBN 0-9728422-1-7, pp. 266-269.
6. Y. Cheng, et al., "BSIM3v3 Manual", UC Berkeley, 1997
7. L. Larcher, P. Pavan, S. Pietri, L. Albani, A. Marmiroli, "A new compact DC model of floating gate memory cells without capacitive coupling coefficients", IEEE Trans. Electron Dev., Vol. 49 (2), pp. 301-307, 2002.
8. The MOS model, level 1101 - http://www.semiconductors.philips.com/Philips_Models/
9. 10.W. Liu, X. Jin, K. M. Cao, an C. Hu, "BSIM4.0.0 MOSFET Model User's Manual," 2000.
10. 11.M. Bucher, C. Lallament, C. Enz, F. Theodoloz, and F. Krummenacher, "The EPFL-EKV MOSFET model equations for circuit simulation, Version 2.6," Technical Report, Electronics Laboratory, Swiss Federal Institute o Technology (EPFL), June 1997. [available online: http://legwww.epfl.ch/ekv/]
11. G. Gildenblat, N. Arora, R. Sung, and P. Bendix, "Scalable surface potential based compact MOSFET model," Proc. 1997 International Semiconductor Device Research Symposium, p. 33, 1997.
12. http://home.hiroshima-u.ac.jp/usdl/HiSIM.html
13. B. Prince, "Emerging memories: technologies and trends", Kluwer Academic Publ., 2002, ISBN 0-7923-7684-6.
14. P. Cappelletti, C. Golla, P. Olivo, E. Zanoni, "Flash Memories", Kluwer Academic Publ., 1999, ISBN 0-7923-8487-3.

15. Frohman-Bentchkowsky D., "A fully decode 2048-bit electrically programmable MOS-ROM", IEEE ISSCC Tech. Digest, p.80, 1971

16. P. Pavan, R. Bez, P. Olivo, E. Zanoni, "Flash memory cells – an overview", Proceedings of the IEEE, Vol. 85, N. 8, Aug. 1997, pp. 1248-1271

17. IEEE Standards Department (1998) "IEEE P1005 draft standar for definitions, symbols, and characteristics of floating gate memory arrays" (approved 1998) IEEE, 44 Hoes Lane, Piscataway, NJ, USA

18. F. Berenga, V. Daniele, G. Ferla, G. Torelli, W. Mienner, I. Haraszti, P. Sieber: "E2PROM TV synthesizer" - 1978 IEEE International Solid-State Circuits Conference (ISSCC), Digest of Technical Papers, San Francisco (California, U.S.A.), Feb. 1978, pp. 196-197.

19. G. Torelli: "An LSI technology fully compatible EPROM cell" - Alta Frequenza, Vol. LI, No. 6, Nov./Dec. 1982, pp. 345-351.

20. G. Torelli, G. Caironi, G. Salomone: "Integrating non-volatile station memory and remote control receiver in a single chip" - 1983 IEEE International Conference on Consumer Electronics (ICCE), Digest of Technical Papers, Des Plaines (Illinois, U.S.A.), June 1983, pp. 12-13.

21. G. Caironi, G. Torelli, D. Devecchi: "A TV frequency synthesis system with a single chip microprocessor interface including non volatile memory" – IEEE Transactions on Consumer Electronics, Vol. 28, No. 3, Aug. 1982, pp. 363-372.

22. G. Torelli, P. Lupi: "An improved method for programming a word-erasable EEPROM" - Alta Frequenza, Vol. LII, No. 6, Nov./Dec. 1983, pp. 487-494.

23. V. Daniele, G. Corda, A. Magrucci, G. Torelli: "Metodo di programmazione per una memoria a semiconduttore non volatile elettricamente alterabile" - Dep. 28/9/1979, n. 26071 A/79. Italian Patent n. 1224062; U.S. Patent 4,357,685, issued 2/11/82.

24. V. Daniele, G. Corda, A. Magrucci, G. Torelli: "Metodo di programmazione per una memoria a semiconduttore non volatile elettricamente alterabile del tipo cancellabile per gruppi di celle" - Dep. 8/10/1979, n. 26306 A/79. Italian Patent n. 1209430; UK Patent GB 2 061 651 B, issued 8/2/84.

Chapter 2

PRINCIPLES OF FLOATING GATE DEVICES
Basic process, operation, physical aspects and reliability

The floating gate device is the basic building block for many types of nonvolatile memories: Flash, EPROM and EEPROM. In a memory product, single FG devices have to be connected together and compacted to use the smaller possible area on silicon. Depending on applications, different architectures have been introduced and manufactured. Some allow parallel access (program and read of a randomly addressed cell) and are better suited for embedded applications, others allow serial access (read and program are performed by page, i.e. more cells at the same time) and are better suited for mass storage applications.

In this Chapter we will illustrate the physical mechanisms involved in FG program and erase operations. We will investigate the effects of these mechanisms on the reliability of the single FG device. We will also show that when a single FG device is part of an array, many reliability issues can arise. All this information will be used to forecast scaling issues of the FG device.

1. TECHNOLOGY HIGHLIGHTS

1.1 Introduction

"The continuous shrink of dimensions…": how many papers start with this or an equivalent statement! In facts the most powerful driving force of the semiconductor industry is the reduction of devices' dimensions. Such shrink has several effects: 1) the reduction of weight and dimensions of the final equipments: PCs, cellular phones, etc…; 2) the possibility of manufacturing new solutions such as Personal Digital Assistants integrated with cellular phones; 3) lower manufacturing costs: mainly due to a higher throughput of production equipments: steppers, etchers, furnaces, etc.; 4) lower power consumption; 5) faster devices.

As clarified by all the papers starting with the sentence reported above, a continuously increasing number of problems arise because of this rush towards the minimum possible dimensions. Probably the most evident is the class of cleanness necessary to manufacture today's integrated circuit: most advanced production lines are now, year 2003, of class 1, meaning that no more than one particle having a linear dimension larger than 0.5 micron can be present in a cube meter. This level of cleanness calls for a sophisticated system of air purification, airflow (laminar flow to avoid turbulence), air pressure (higher with respect to the outside environment so that the dirt external air is continuously pushed out of the clean area). Moreover, recently, to avoid even more the contamination with the dirt external ambient, a new way to manage the wafers has been developed: in the most advanced fabrication lines, wafers are always kept within their box when they are not being processed. When it is their time to be processed, the box is placed in front of the equipment, it is loaded, then it is opened inside the equipment and then he wafers are ready to be processed.

But this is just the most evident aspect; all the technological steps have to concur to this reduction of dimensions: lithography, cleanings, insulation technologies, ion implantation, silicon oxidation, doping activation, etching, deposition.

One further aspect which is worth mentioning is the overall uniformity control of the technological steps. We mean both the random and the systematic variations within a single die and die-to-die (including wafer-to-wafer and lot-to-lot effects). All the systematic effects have to be corrected or taken into account. For instance the dimensions of an isolated line can be different from the dimension of the same line in a region dense of structures: this could be taken into account changing the masks dimensions. Another example regards the layers' thickness variation along the wafer's radius: the process has to be optimized to avoid these variations.

To clarify the effect of these variations, let's consider the channel length of memory cells: the designed channel length will have to take in consideration that of the more than half a billion cells of a 512 megabit circuit, the shortest and the longest cell will have to work properly during reading, erasing and programming operations. So, both circuit design and process architecture design will have to cope with these effects, in general increasing the final circuit's dimension or worsening its performances. The process steps need to be optimized in order to reduce as much as possible such variations.

Moreover, as we will try to demonstrate, more and more the interactions among the different process steps are playing a role in the evolution of silicon technologies. Among all these steps, lithography has always been the bottleneck for the shrinkage of devices. No surprise, then, realizing that lithography is the most expensive technological step; its cost is about one third of the overall manufacturing costs. A lithography platform for 193 nm exposure is more than 15 million dollars, while a set of masks in 90 nm technology for Flash EEPROM production is approaching the million dollars.

1.2 Lithography

Two main technological aspects are related to lithography: the alignment among mask levels, and the resolution, i.e. the minimum dimension that can be patterned.

1) Alignment. The capability of the exposing machine to align masks has a direct impact on the distances and hence on the final dimensions. As an example let's consider the layout rules of a memory cell in an "old" technology. The dimension (d) of the cell in one of the two directions is:

$$d = c_{d/2} + c\_g_o + c\_g_d + g_d + s_{d/2} \tag{1}$$

where o refers to overlay and d to dimensions, and the symbols are explained in the following table. Typical values for the considered technology are:

Table 2-1. Cell dimensions in "old" technology.

Issue		Short	Value
Half contact	dimension	$c_{d/2}$	110 nm
Contact to gate	overlay	c_g_o	45 nm
Contact to gate	dimension	c_g_d	75 nm
Gate length	dimension	g_d	300 nm
Half source	dimension	g_d	110 nm

This means that the "alignment" needs 45 nm over 620 nm, i.e. more than 7%. With a perfect alignment one could gain around the 14% (7% in the two directions) in area.

A way to improve the alignment in a production line is to use for the same lot of wafers the same piece of equipment to expose the critical masks. The cost of this is so high in terms of cycle time and equipment exploitation, that this solution is very seldom applied.

Another not easy solution that is always pursued in process development is the "self alignment": by this we mean technological steps for which the levels are by construction aligned each other not thanks to exposure. The simplest example is the source and drain implants of MOS transistors: the alignment to the gate is obtained thanks to the gate itself that prevents the dopants to reach the channel and allows the doping of the source and drain regions. This "self alignment" has been more and more extended to other process modules.

2) Resolution. The resolution of an exposing machine (RES) is proportional to the used wavelength (λ) and is inversely proportional to the numerical aperture (NA) of the system:

$$RES = R_1 \lambda / NA \qquad (2)$$

The ways to improve the resolution are the following: 1) reduction of the wavelength, 2) increase of the NA, 3) reduction of the proportionality coefficient R_1.

Starting from the wavelength, currently a wavelength of 248 nm is used for mass production of 130 nm technology. Already a tremendous result has been obtained with respect to the past: the minimum dimension is one half of the used wavelength. For the 90 nm technological node, the 193 nm wavelength, provided by an excimer Krypton-Fluorine (KrF) laser, is used. To further reduce this wavelength it will be necessary to substantially modify the masks and the exposing machines. In facts the quartz is opaque to the following wavelength node (157 nm), and hence it cannot be used anymore for masks and for lenses. Likely the solution will be the use of Carbon di-Fluorine lenses together with a catadiotric system [1].

As already noted the R_1/NA ratio can be of the order of 0.5, as we can pattern structures as wide as half the wavelength. The Numerical Aperture is the sine of the maximum incident angle. To increase this angle is necessary to use larger lenses keeping the same level of distortion, and, for very high angles, to use immersion lithography, with the relative costs and problems. Besides the costs of the exposing machine, the increase of NA reduces the depth of focus of the lithographic process, thus calling for a better planarization that will be discussed further on.

The last way to improve resolution is the reduction of R_1. Different interesting techniques have been developed to reduce this parameter. One of the first is the Phase Shift Mask (PSM) process (see Fig. 2-1). One additional layer has to be used for the mask production, and hence the costs increase. As illustrated in the figure, the shifter allows a destructive interference in the region between two openings in the chrome layer region. Another set of techniques to improve the resulting image consist in modifying the pattern on the mask in order to exploit diffraction effects.

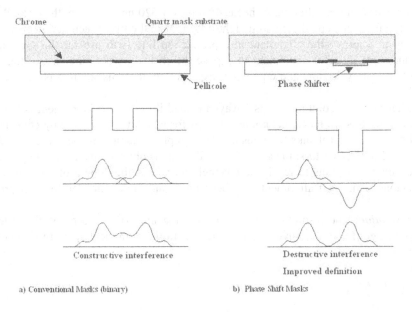

Figure 2-1. Conventional Masks vs Shift Phase Masks.

As we said, increasing the NA results in a decrease of the depth of focus, thus it is necessary to have all the structures to be exposed as close as possible to a common plane. To reach this objective different planarization techniques have been developed. Currently the most used is the "Chemical Mechanical Polishing" (CMP). As evident from the acronym, this is a Chemical (through an etching component) Mechanical (through a brush) technique to planarize the surface of the wafer. Another technique in use is the "Shallow Trench Isolation" (STI) which is a technique to electrically isolate active devices by means of silicon dioxide: it is obtained etching a trench in silicon and filling it with silicon dioxide. The resulting structures is much more planar with respect to other techniques, as a hole in silicon is completely filled till the top. The advantage with respect to previous technologies consists in more planar structures as a result of the isolation module.

1.3   Field isolation

After lithography, the most critical process step is the already mentioned isolation of active components. The main isolation is performed via silicon dioxide. As always in microelectronics, the main goal is the reduction of dimensions. As far as the isolation is concerned, this reduction is obtained mainly via a sharp transition between active and isolation region and via the realization of a structure as planar as possible, in order to facilitate the lithography steps, as mentioned above. The evolution of this process steps has been in these two directions. The first isolation schemes, named LOCOS (LOCal Oxidation of Silicon) consisted of a field oxide obtained oxidizing the planar surface of silicon and masking the active region with a sandwich of silicon dioxide + silicon nitride. This was followed by the Polycrystalline-silicon Buffered LOCOS (PBL), a similar scheme but with a sandwich of silicon dioxide + polycrystalline silicon + silicon nitride, designed in order to reduce the transition between isolation and active region. The following processes made use of a recession of the silicon dioxide, obtained via an etch of the silicon region to be oxidized, in order to obtain a more planar structure (see Fig. 2-2); this technology was named recessed LOCOS.

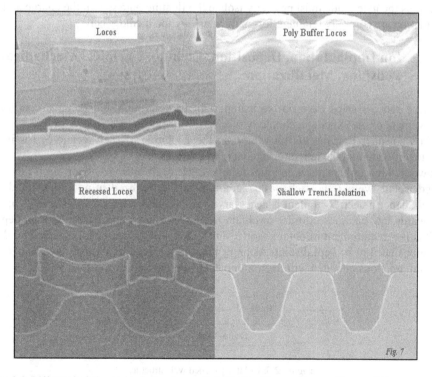

Figure 2-2. Different isolation technologies: Locos, Poly Buffered Locos, Recessed Locos, STI.

The currently used technology is the Shallow Trench Isolation (STI): in the isolation region the silicon is etched, and the resulting structure is filled with deposited silicon dioxide. Besides the planarization, which is common to the recessed LOCOS, a sharper transition between isolation and active region is obtained (see Fig. 2-2). For the 90 nm Flash process the minimum pitch of the isolation structure is 100 nm + 100 nm [2].

The most critical aspect of the isolation step is the stress induced in active region of the device. The strain caused by this stress is responsible for the behavior of the active device: it has been shown that the current driven by the MOS transistor is strongly dependent on the active area dimensions [3].

1.4   Silicon oxidation

Silicon oxidation deserve a dedicated section as this is, in particular for floating gate devices, one of the most critical process steps. This process starts with the cleaning of the silicon surface to prepare the following oxidation. Oxide growth is crucial for a reliable dielectric layer. For MOS transistors, the gate oxide determines the coupling between gate and channel and allows the isolation of the gate electrode. Even more for a FG device: the gate oxide has to behave at the same time as a "conductor" and as an "insulator". The programming and erasing operation rely on the capability of the silicon dioxide to allow the flow of electrons. On the other side during reading and in all other conditions the gate oxide must perfectly isolate the floating gate so that the charge and the information are correctly retained, for many years, by the memory cell.

1.5   Ion Implantation, Deposition, Etching, Chemical Mechanical Polishing, Metallization

These process steps have not seen important breakthrough, but rather a continuous improvement of the existing technology.

Let's consider first the *ion implantation*: the evolution of such technology is in the direction of widening the range of implantation energies:
- towards high implantation energies (above 2 MeV) in order to obtain deep junctions and isolate wells (e.g. each flash memory sector is realized in an isolated P-doped well, hence a deep N-implant is necessary to isolate such P-doped wells (Fig 2-3);
- towards low energies in order to obtain shallower Source and Drain junction depths. This is a requirement to obtain shorter and shorter MOS transistors: source and drain regions need to be kept distant. As an example, P-channel transistors have source and drain junctions obtained with Boron implants with energies lower than 1 KeV.

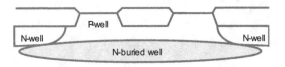

Figure 2-3. Isolated p-doped well structure.

For the sake of completeness, another trend in ion implantation technology is the use of tilted implantation in order to obtain pockets to reduce short channel effects. For this reason all processes are getting colder: source and drain doping species must not contribute to reduce the channel region.

The main aspects related to *deposition* and *etch* process steps are the cleanness and the uniformity across the wafer and between wafers and lots. In today Flash memory processes the insulator between the floating gate and the control gate is a sandwich of silicon dioxide, silicon nitride and silicon dioxide. Till $0.5 - 0.35$ μm generations, oxide layers were obtained by oxidizing the polycrystalline silicon and the silicon nitride layers because the obtained oxide had a better quality compared to the deposited silicon dioxide. As mentioned the need of "cold" processes pushed in the direction of layer depositions. Once again the uniformity is a critical issue, both within the single cell and between cells: so both the layer thickness and the gate length have to vary as slowly as possible throughout the dice.

The objective of *CMP* is the planarization of the structures. As we have explained above, this is fundamental because it allows a smaller depth of focus for the mask exposure, that in turns allows a better resolution of the lithography process. But not only this, e.g. the deposition of metal layers on abrupt steps can cause the crack of the same metal layer.

The last process step we want to mention is *metallization*. The major change of the last years is the introduction of copper as interconnection layer. The advantage of copper with respect to aluminum is the lower resistivity and the higher immunity to electromigration. The introduction of copper and of low-k dielectric layers results in a reduction of the interconnect capacitance in the range of about 30-40%. These new materials make it necessary to use a new patterning technique that is the damascene in simple and double version.

In the simple version, after the contacts have been filled, usually with tungsten, the dielectric between polycrystalline silicon and 1st metal level is deposited and is patterned in the region where the copper is expected to remain. Then copper is deposited and by CMP is removed everywhere except in the mentioned pattern region (see Fig. 2-4).

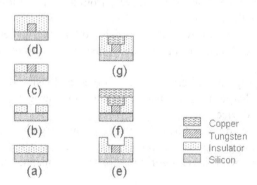

Figure 2-4. The damascene process: a) oxide deposition; b) contact definition; c) tungsten filling; d) oxide deposition; e) "damascene etch"; f) copper deposition; g) copper CMP.

The double damascene allows having both contacts and layer of copper. Let's suppose to apply this methodology after the CMP step mentioned above, an intermetal dielectric is deposited. A first masking step etches this layer in the contacts region, then a second masking step etches, as in the single damascene the region where the copper is to remain resulting in the situation of Fig. 2-5c. Also in this case copper is deposited everywhere and then removed by CMP resulting in the structure of Fig. 2-5e.

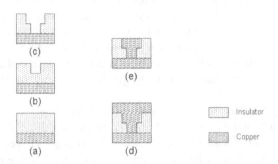

Figure 2-5. The dual damascene process: a) oxide deposition; b)contact "damascene" definition; c) "damascene second etch"; d) copper deposition; e) copper CMP.

2.   CELL OPERATION

In this section we will introduce the principles of the physics involved in the program and erase operations of the single FG device.

As already stated, to have a memory cell that can commute from one state to the other (from "0" to "1") and that can store the information independently of external conditions, the storing element needs to be a device whose conductivity can be changed in a nondestructive way. The FG memory device is a particular kind of MOS transistor whose threshold voltage that can be changed repetitively from a high to a low state by changing the amount of charge in the floating gate.

Read operation is performed by applying a gate voltage that is between the values of erased and programmed threshold voltages and sensing the current flowing through the device.

Write operations (program and erase) are performed by biasing the FG device in such conditions to allow charge transfer from/into the floating gate, thus changing the conductivity of the FG device.

Memory products are composed by cell arrays and ancillary circuits. Ancillary circuits are used to address the array, to manage power supply; in new generations there is also a microprocessor to implement some algorithms for "smart" management of the memory array.

2.1   Charge injection mechanisms

Many solutions have been used to transfer electric charge from and into the FG. Both for erase and program, the problem consists in making the charge pass through a layer of insulating material (silicon dioxide).

The Channel Hot Electron (CHE) injection mechanism is generally used to program Flash memories: a lateral electric field (between source and drain) "heats" the electrons and a transversal electric field (between channel and control gate) promotes the injection of the hot carriers in the FG. Many Flash memory generations used CHE program mechanism. This is a power consuming mechanisms, due to large currents and low injection efficiency, but it has the advantage of an easy control of the amount of transferred charge. Therefore, programming the whole array results in very narrow threshold voltage distributions. In any case, the microprocessor in the memory device implements a Program and Verify algorithm to control the threshold voltage of the single memory device in the array.

In the last years, a new Flash memory program mechanism, namely the CHannel Initiated Secondary ELectron (CHISEL), has been proposed to overcome the major disadvantages of CHE injection: mainly, the large voltage required and the large current involved. This mechanism is based on "hot carriers" phenomena and exploits the negative biasing of the body to increase the efficiency of the electron injection into the FG [4-6]. In the CHISEL ionization feedback process, electrons travel from source to drain, are heated by the lateral field, and impact-ionize at the drain. Generated electrons travel to the gate, while holes are collected at the substrate. While traveling to the substrate, also secondary holes ionize, generating electron/hole pairs. The so-generated electrons move toward the floating gate gaining energy from the silicon field and reaching the silicon/oxide interface where the higher oxide field is more favorable for the electron injection. Therefore, the CHISEL programming efficiency is increased compared to the CHE one.

To increase the bandwidth, the programming speed of a single cell and the parallelism of the programming operation need to be increased. To increase programming speed, programming current needs to be augmented, but this is in contrast with limitations from charge pumps.

New Flash generations will be programmed using Fowler Nordheim (FN) tunneling currents. The FN tunneling mechanism is based on the electron tunneling through a thin oxide barrier, whose probability rises exponentially with the electric field, determining a non negligible tunnel current in high field conditions. FN currents are much smaller than CHE, but require high voltages. The FN tunneling mechanism is also used to erase the cell.

Unfortunately, the threshold voltage distribution of cells programmed or erased by FN is broader. Thus, constraints on erase threshold distribution or on program threshold distribution become tighter and tighter and new studies on these topics are required to optimize new families of Flash memories.

2.2   Channel Hot Electron current

CHE injection current is the most used mechanism to program Flash memories. Furthermore, hot carrier phenomena degrade significantly MOS and NVM device reliability and performances. Therefore, following advances in device concepts and technology, the interest in studying hot carrier injection current has grown considerably in the last years.

Detrimental effects of hot-electrons (namely: drain current reduction, small signal performance degradation, threshold voltage shift and I_D sub-threshold slope lowering) are due to both interface state generation and the charge trapping in the portion of oxide above the drain junction. Thus, to gain insights into degradation of both MOS performances [7-9] and Flash memory cell reliability [10-11] hot carrier phenomena have been studied through experimental techniques. Moreover, the CHE current has been studied also by means of models developed to reproduce real features of the phenomenon.

Unfortunately, since physical mechanisms of the CHE current are quite complex, the correct modeling of this current is a very difficult issue. For this reason, this task has been mostly tackled using 2-D device simulators, which incorporated models of the CHE currents [12-13]. Nevertheless, also CMs of CHE current have been proposed in the literature [13-15].

The physical mechanism of CHE injection is relatively simple to understand qualitatively. An electron traveling from the source to the drain gains energy from the lateral electric field and looses energy to the lattice vibrations (acoustic and optical phonons). At low fields, this is a dynamic equilibrium condition, which holds until the field strength reaches approximately 100 KV/cm [16]. For fields exceeding this value, electrons are no longer in equilibrium with the lattice, and their energy begins to increase. Electrons are "heated" by the high lateral electric field and a small fraction of them gains enough energy to surmount the oxide barrier. For an electron to overcome this potential barrier, it must have kinetic energy higher than the potential barrier and velocity directed towards the FG [15].

To evaluate how many electrons will actually cross the barrier, one should know: *i)* the energy distribution as a function of lateral field; *ii)* the momentum distribution as a function of electron energy; *iii)* the effective height of the oxide potential barrier; *iv)* the probability of an electron with energy E, wave vector k, and distance d from the Si/SiO$_2$ interface to cross the barrier. Moreover, when the energy gained by electron reach a threshold, impact ionization becomes a second important energy loss mechanism which needs to be included [17].

A description of the injection conditions can be accomplished with different approaches.

The CHE injection current is explained and often simulated following the "lucky electron" model [14]. This model is based on the probability for an electron to be lucky enough to travel ballistically for a distance several times the mean free path without scattering, eventually acquiring enough energy to cross the potential barrier if a collision pushes it towards the Si/SiO$_2$ interface. Consequently, the probability of injection is the lumped probability of the following statistically independent events: 1) the carrier is "lucky" enough to acquire the energy to overcome the oxide barrier and to retain this energy after the collision that redirects it towards the interface; 2) the carrier does not suffer collision traveling the distance to the interface and surmounting the repulsive oxide field at the injection point across the barrier [18]. Although this simple model does not fit precisely with some experiments, it allowed a straightforward and quite successful simulation of the experimental gate current, and it can be included in Spice-like model of MOSFET transistors and FG memory cells.

A more rigorous model of CHE current is based on the quasi-thermal equilibrium approach [19-20]. It assumes that the electron can be treated as a gas in quasi-thermal equilibrium with the electric field. This electron gas is characterized by an "effective temperature", which is different from the lattice temperature. The model establishes a non-local relation between the effective electron temperature and the drift-field [21], thus the carrier probability to acquire certain energies depends on the complete profile of the electric field in the channel region [19].

Both models allow the prediction of the correct relation between the gate and the substrate current [18], which is composed of holes generated by impact-ionization in the drain region.

2.3 CHannel Initiated Secondary ELectron current

The CHannel Initiated Secondary ELectron (CHISEL) mechanism has been introduced in the last years to improve the efficiency of Flash memory program operation. In fact, the CHISEL mechanism overcomes the major disadvantages of CHE injection, that are the large voltage and the large current [4-6, 22].

The detailed analysis of complex physical phenomena beneath such a program mechanism requires the use of sophisticated Monte-Carlo transport simulations that are necessary to model accurately the high energy tail of the energy distribution of carriers involved in impact ionization phenomena [4].

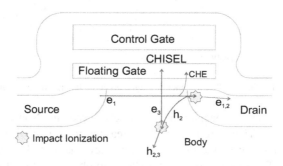

Figure 2-6. Schematic cross section of a Flash memory cell, where physical phenomena involved in CHISEL injection mechanism are evidenced.

Nevertheless, a qualitative understanding of CHISEL injection is easy to achieve looking at the physical mechanism sketched in Figure 2-6.

Schematically, channel electrons, e_1, injected into the drain ionize producing electron-hole (e_2-h_2) pairs: e_2 are collected at the drain, whereas h_2, heated by the high electric field at the drain junction, ionize again. Thus, new electron-hole (e_3-h_3) pairs are generated: holes h_3 leave through the substrate, whereas e_3 electrons are driven toward the Si/SiO$_2$ interface, reaching it where the oxide field is more favorable to cross the barrier.

2.4 FOWLER-NORDHEIM TUNNELING CURRENT

The solution of the Schroedinger equation shows that a tunneling through a potential barrier is possible even for classically forbidden barriers [23], and applies well to MOS structures with thin oxide. Figure 2-7 shows the energy-band diagram of a MOS structure with negative bias applied to the n+poly gate with respect to the p-doped silicon substrate.

The probability of electron-tunneling depends both on the distribution of occupied states in the injecting material and on the shape, height and width of the potential barrier.

Using a free-electron gas model for the metal and the Wentzel-Kramers-Brillouin (WKB) approximation for tunneling probability [24], the classical expression of the Fowler-Nordheim (FN) current density can be obtained, which will be widely discussed in Chapter4 [25].

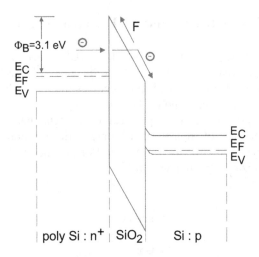

Figure 2-7. Fowler-Nordheim tunnel through a potential barrier in a MOS structure.

The tunneling-injection mechanism is widely used in Non-Volatile Memories (EEPROM and Flash). There are mainly three reasons for this choice: i) tunneling is a pure electrical mechanism; ii) the involved current level is quite low, thus it allows the internal generation of supply voltages; iii) it allows to obtain the time to program (<1ms) twelve orders of magnitude shorter than retention-time (>10 years), which is a fundamental request for all NVM technologies.

On the other hand, the exponential dependence of FN tunnel current on the oxide field causes some critical problems of process control, because even a very small variation of oxide thickness among the cells in a memory array results in a great difference in programming or erasing currents, thus spreading the threshold voltage distribution.

The optimum tunnel oxide thickness for FG memories based on tunneling phenomena is chosen trading off between performances constraints (programming speed, power consumption, ...) which would require thin oxides, and reliability concerns, which would require thick oxides.

In fact, tunneling currents are also important for device-reliability at low fields. In the case of low quality tunnel oxides, or when oxides are stressed many times at high voltages/currents [26], Trap Assisted-Tunneling (TAT) through bulk traps either present or generated in the oxide can strongly enhance the tunnel current, thus threatening severely NVM data retention requirements. Therefore, oxide defects must be avoided to control program/erase characteristics and to obtain good reliability.

Finally, although the classic form of FN current fits quite well experimental data, many features have been still undervalued: the temperature dependence of the phenomenon; the correct electron statistics, Fermi-Dirac and not Maxwellian [27]; and the quantum-effects at the silicon-oxide interface. When the silicon surface is inverted or accumulated (which are usual conditions during tunneling injection), carriers are confined into a narrow potential well, so that their energy is quantized. In modern VLSI-USLI devices, the electron energy quantization cannot be neglected for a proper modeling of the tunnel current, and therefore its main effects on the FN current have to be considered, that are: the lowering of the oxide barrier height, that becomes oxide field dependent, and the reduction of the oxide field compared to that classically calculated [28].

3. DISTURBS AND RELIABILITY

Among the critical problems affecting memory circuits we now focus on array disturbs and erase-threshold voltage distribution. These are routinely tested in manufactured devices. The simplest array is the NOR array, Fig. 2-8, where cells which are on the same column share the same source/drain diffusion (Bit Line, BL) and cells on the same row are connected through the same polycrystalline silicon line which forms the gates of devices (Word Line, WL). The information stored in each single device can be accessed (read or written) by applying the correct voltages to the BLs and WLs of the memory array. It is important to stress how the cell considered as a single device, and the cell as an element of the array are very different. Access to the single array element is now performed through other elements of the same array, either on the same column or on the same row. Moreover, the operations to be performed on a single element of the array have now an influence on the elements of the same row/column.

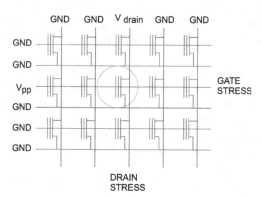

Figure 2-8. NOR Array. When the circled cell is selected, cells in the same row/column undergo gate stress/drain stress.

The confidence in nonvolatile memory reliability has grown together with the understanding of memory-cell failure mechanisms. Cycling and retention experiments are performed to investigate single-cell reliability. The high degree of testability allows the detection at wafer level of latent defects, which may cause single bit failures related to programming disturbs, data retention, and premature oxide breakdown. For this reason, for example, Flash memories are more reliable than full-featured EEPROMs at equivalent density [29].

Starting from this industry standard T-shaped cell layout connected in a NOR array, new solutions have been developed to scale dimensions of the final chip. AMG [30], for example, accomplishes cell size scaling by sharing one metal line per two diffusion bit-lines; a new segmentation scheme and fieldless array allow the achievement of the minimum design rule of the process, which typically is the pitch of the polycrystalline silicon.

3.1   Programming Disturbs

Consider an array of FG devices as in Fig. 2-8. To program the highlighted cell, a high voltage (V_{pp} around 9V) is applied to the WL and a sufficiently high voltage (V_{drain} around 4.5-5V) is applied to the BL to generate hot electrons to program the cell. In this bias condition, though, there are two major disturbs, one due to the high voltage applied to the WL, the second to the medium – high voltage applied to the BL.

High voltages applied to the WL can stress the memory cells that have their gate connected to the WL but are not selected. There might be tunneling of electrons from the FG to the control gate through the interpoly oxide in all the programmed cells, i.e., in those cells where the FG is filled with electrons, since they have, for example, 9V applied to the gate and 0V on both source and drain. This is the *dc-erasing* disturb, which induces charge loss and therefore reduces the margin for the high level of threshold voltage. There might be also tunneling of electrons from the substrate to the FG in all the non-programmed cells, i.e., in those cells where the FG is "empty." This is the *dc programming* disturb, which induces charge-gain and reduces the margin for the low level of threshold voltage. Both of these disturbs are called *gate disturbs* and are present even during reading operations. They are triggered during fabrication testing to assess the gate-oxide quality.

A relatively high voltage applied to the BL can stress the drains of all FG transistors in the same column. Namely, in cells which share the BL with cells which are to be programmed, electrons tunnel from the FG through the gate oxide to the drain [31]. Moreover, holes can be generated via impact-ionization in the substrate and then injected in the FG. This disturb, called *drain disturb*, causes charge loss and, consequently, a decrease in the high value of the threshold voltage. The same disturb can result from extensive reading cycles and can be used as a gate oxide quality monitor.

These disturbs become important when the same reading or programming operation are repeated continuously, for example, when a complete row or column is programmed in an array. In a 1-Mb array, this requires a thousand repetitions. Disturb influence becomes more and more important on increasing the number of reading-programming or programming-erasing cycles.

3.2   Retention

Fast program and erase operations require high voltages and currents through thin oxides, which in turn are easily degraded. In modern Flash memory cells, FG capacitance is approximately 1 fF. A loss of only 1 fC can cause a 1-V threshold voltage shift. If we consider the constraints on data retention in ten years, this means that a loss of less than five electrons per day can be tolerated.

Mechanisms that lead to charge loss or charge gain can be divided in two categories: extrinsic and intrinsic. The former are due to defects in the device structure; the latter are due to the physical mechanisms that are used for program and erase operations.

a) Intrinsic mechanisms. Intrinsic mechanisms that contribute to charge variations are field-assisted electron emission, thermoionic emission, and electron detrapping. The first mechanism, field-assisted electron emission [32], consists of the motion of electrons stored in the FG of a programmed cell, which can migrate to the interface with the oxide and from there tunnel into the substrate, thus causing charge loss. If the cell is erased, i.e., has a low threshold voltage, the opposite injection can happen.

Experiments have demonstrated that the leakage current due to these mechanisms depends on the floating-to-control-gate coupling coefficient α_G and on the stress level [32]. The probability of an electron passing through the oxide barrier due to tunneling depends upon the voltage drop between the FG and substrate. Thus, being the FG potential dependent on the control-gate potential through α_G, the charge loss depends on α_G. The leakage current depends exponentially on the electric field (it is a tunneling phenomenon), and the electric field around the FG can be approximately calculated through:

$$E = \frac{Q}{2\varepsilon_{ox}\sigma} \tag{3}$$

where ε_{ox} is the silicon dioxide dielectric constant and σ is the FG area. If Q decreases, the electric field vanishes and as a consequence the leakage current decreases. Moreover, field reduction reduces also some other degradation processes which are field enhanced (e.g. drift diffusion of mobile ions). The second mechanism of charge loss, the thermoionic emission, is a mechanism of emission of carriers above the potential barrier. At room temperature this phenomenon is negligible, but it becomes relevant at high temperatures [33]. Last, detrapping of electrons in the gate oxide is a charge loss mechanism that reduces the program threshold voltage.

b) Extrinsic causes. Extrinsic causes that can influence the charge storage are oxide defects and ionic contamination. Oxide defects can cause charge loss or gain [33]. If the cell is programmed, with no bias applied, its FG has a negative potential due to the stored charge. This potential induces an electric field in the oxide surrounding the FG itself, which can be as high as 1-2 MV/cm. Therefore, defects present in the oxide layer can form conductive paths, which increase significantly the current at low fields and tend to program the cell. If the memory cell is overerased and stores a positive charge, the electric field will induce a charge gain.

Ionic contamination is a big problem in every nonvolatile memory technology [31,33,34]. Ions, usually positive ones, are attracted to the FG which is negatively charged, thus shielding its effects and inducing charge loss. Memory chips can be affected by contaminations, which, during passivation deposition, can penetrate through defects in passivation glasses or from chip edges. The quality of passivation layers has to be increased in order to reduce this effect. Retention capability of Flash memories has to be checked by using accelerated tests, which usually adopt high electric fields and hostile environments at high temperatures.

3.3   Endurance

Cycling is known to cause a fairly uniform wear-out of cell performance, which eventually limits memory endurance. Experiments to test endurance are performed applying constant program/erase pulses. The variations of program and erase threshold levels give a measure of oxide aging. In real devices, this corresponds to longer program/erase times.

The quality of tunnel oxide is again crucial for reliability assessment.

3.4   Erase Distribution

In Flash memory integrated circuits, the complete erase operation is indeed a sequence of elementary erase operations. A first erase pulse is generated internally and sent to the logic circuitry, which controls the erase operation and is integrated in the same chip with the memory array itself. An algorithm controls whether or not the erase operation is completed.

The verification of the complete erasure of all the cells in a block, which can be very large, is one of the biggest issues in Flash technology. After erase the values of the threshold voltage values of the erased cells are checked: these values have a distribution that is specific to the process. This distribution spreads around an average value and needs to be easily controlled for each process and to have a small variance. Fig. 2-9 shows the threshold voltage distribution for a 1-Mb Flash device [29]. The distribution seems to be Gaussian, but it is not symmetrical toward lower values. As can be seen, a high percentage of the cells have a very small variation of the threshold voltage, and only a very small percentage show very large threshold variations. This very small percentage has a great relevance. It is used as a process monitor since it is the limiting factor for the whole Flash technology. In fact, a memory cell with negative or zero threshold voltage is on (it drives current) also when the control gate voltage is zero, i.e. the cell is not selected. Thus, since reading a single cell in the array means to sense all the currents driven by the memory cells in the same column (as there is no selection transistor), this causes a logic error, as the overerased (unselected) cell supplies current independently on the erased/programmed state of the cell selected to be read.

The exponential tail in threshold voltage distribution (see Fig. 2-9) represents a large population of cells that erase faster than typical bits. This population is too large to be attributed to extrinsic defects, and it is believed to be related to statistical fluctuations of oxide charge and to the structure of the injecting electrode [29]. Positive charges in the tunnel oxide and irregular polycrystalline grains may induce a local increase of the electric field, thus enhancing current injection locally and making individual cells erase faster than average. This explanation is consistent with the observation of the narrowing of the tail on increasing the number of program/erase cycles performed on the array. In fact, a larger current density corresponds to a higher negative trap generation rate, resulting in faster aging. The generated negative charge partially neutralizes electric field peaks, making the current injection more uniform and the erasing speed of tail bits closer to that of typical bits [29].

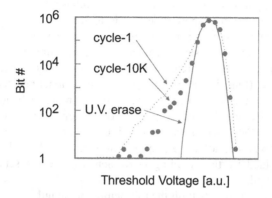

Threshold Voltage [a.u.]

Figure 2-9. Erase Threshold Voltage distribution of 1MB array; UV erase distribution is also shown for comparison with electrical erase after 1 program and 10K P/E cycles.

Other failures are related to the erase mechanism. Since FN tunneling is not self-limiting, it can lead to overerasing of the memory cells, i.e., more electrons than those which have been trapped are removed from the FG. The device has less negative charges than in the nonprogrammed case and a net positive charge is now present, thus transforming the device from an enhancement to a depletion device.

3.5 Scaling issues

The architecture of an industry-standard Flash-cell array is typical of a NOR gate array, where every single cell is addressed by two signals, one for the BL and one for the WL; the source line and body are common to the whole array. Moreover, in standard arrays, a contact is shared between two cells, thus consuming a lot of cell area.

The common issue among the different solutions and applications is the cost-per-bit reduction, which will be provided mainly by technology scaling.

No consolidated theory has been developed for Flash-cell scaling [35]. Scaling issues deal then with the single cell layout. The goal is to reduce the area used for contacts, and layout issues are contact placement issues. To improve integration, many new solutions have been proposed, mainly new array architectures. A reduction of the area occupied by a Flash memory cell when fabricated in a double-poly stacked gate structure, particularly the reduction of the effective channel length L_{EFF} gives many advantages, not only from the density point of view but also for the performances. In fact, the efficiency of the carrier injection into the FG increases on decreasing L_{EFF}, thus speeding up the program operation. On the contrary, decreasing L_{EFF} enhances punch-through and drain turn on, since the capacitive coupling between the drain and FG increases. The final value of L_{EFF} comes from a tradeoff between performances and disturbs.

Another relevant issue in Flash memories is the need for high voltages for program and erase. While CMOS technology scaling requires the reduction of the operating voltages, the actual program/erase operations are based on physical mechanisms whose major parameters do not scale (3.2-eV energy barrier for CHE and at least 8–9 MV/cm for FN data alteration in 0.1–1 s). Moreover, the trend toward increasing the programming throughput will even force the internal voltage to rise.

When internal generation of power supply is to be done, many issues need to be discussed. For example, internal charge pumping circuits can be used only when small currents flow in the channel, to save area. Erasing opens similar issues. Non-volatility implies at least ten years of charge retention. Non-volatility issues affect the scalability of thin active dielectrics (tunnel and interpoly). A direct tunneling mechanism fixes the tunnel oxide limit to 6 nm, which needs to be increased more realistically up to 7–8 nm due to trap-assisted electron tunneling caused by oxide aging. The scalability limit of the interpoly dielectric (ONO) has been reported to be around 12–13 nm [36]. These thicknesses can be combined to give an equivalent memory cell oxide (defined as tunnel oxide thickness divided by the coupling coefficient α_G), which sets the limit for the memory-cell poly length.

Furthermore, other constraints limit the minimum poly length.
- CHE injection requires some minimum drain-gate overlap and abrupt junction to maximize injection efficiency.
- FN tunneling to the source requires an overlap with the n region below the gate.
- FN tunneling to the channel requires small gate/diffusion overlaps.

When charge is injected from the polycrystalline silicon FG, the number of poly grains in the tunneling area plays an important role in determining the distribution width [37].

In this scenario, the search for higher integration goes toward new architectural solutions, the reduction of the number of contacts, and the reduction of alignment tolerances.

REFERENCES

1. L. Van den Hoven, A. M. Goethals, K. Ronse, M. Van Bavel, and G. Vandenberghe, "Lithography for sub 90 nm applications," in *IEDM Tech. Dig.*, pp.3-8 ,2002.
2. D. Canali, D. Fattori, G. Ginami, G. Girardi, P. Scintu, L. Tarchini, D. Tricarico, "Survey on Flash technology with specific attention to the critical Process Parameters related to the design," in *Proc. of the IEEE*, Vol. 41(4), pp. 503-522, 2003.
3. R. A. Bianchi, G. Bounche, and O. Roux-dit-Buison, "Accurate Modeling of Trench Isolation Induced Stress effects on MOS Electrical Performance," in *IEDM Tech. Dig.*, pp.117-120, 2002.
4. J.D. Bude, M.R. Pinto, and R.K. Smith, "Monte Carlo simulation of the CHISEL-Flash memory cell," *IEEE Trans. Electron Dev.*, Vol. ED-47(10), pp.1873-1881, 2000.
5. J.D. Bude, M. Mastrapasqua, M.R. Pinto, R.W. Gregor, P.J. Kelley, R.A. Kohler, C.W. Leung, Y. Ma, R.J. McPartland, P.K. Roy, R. Singh, "Secondary Electron Flash – a high performance, low power Flash technology for 0.35 µm and below," in *IEDM Tech. Dig.*, pp.279-282, 1997.
6. J. D. Bude, A. Frommer, M. R. Pinto, and G. R. Weber, "EEPROM/flash sub 3.0V drain-source bias hot carrier writing," in *IEDM Tech. Dig.*, pp.989-991, 1995.
7. Y. Leblebici, and S.-M. Kang, "Modeling of nMOS transistors for simulating of hot-carrier-induced device and circuit degradation", *IEEE Trans. on CAD.*, Vol. 11(2), pp.235-246, 1992.
8. P. Chen, L. Wu, G. Zhang, and Z. Liu, "A unified compact scalable ΔI_D model for hot carrier reliability simulation," in Proc. *IEEE 37th Annual International Reliability Physics Symposium*, San Diego, California, 1999, pp. 243-248.

9. C.-M. Yih, S.-M. Cheng, and S.S. Chung, "A new approach to simulating n-MOSFET gate current degradation by including hot-electron induced oxide damage," *IEEE Trans. Electron Dev.*, Vol. ED-45(11), pp.2343-2348, 1998.

10. C. Chen, Z.-Z. Liu, and T.-P. Ma, "Analysis of enhanced hot-carrier effects in scaled Flash memory devices," *IEEE Trans. Electron Dev.*, Vol. ED-45(7), pp.1524-1530, 1998.

11. S.S. Chung, C.-M. Yih, S.-M. Cheng, and M.-S. Liang, "A new technique for hot carrier reliability evaluations of Flash memory cell after long-term program/erase cycle," *IEEE Trans. Electron Dev.*, Vol. ED-46(9), pp.1833-1889, 1999.

12. S. Keeney, R. Bez, D. Cantarelli, F. Piccinini, A. Mathewson, L. Ravazzi, C. Lombardi, "Complete Transient Simulation of Flash EEPROM Devices", *IEEE Trans. Electron Dev.*, Vol. ED-39(12), pp. 2750-2757, 1992.

13. C. Fiegna, F. Venturi, M. Melanotte, E. Sangiorgi, and Bruno Riccò, "Simple and efficient modeling of EPROM writing," *IEEE Trans. Electron Dev.*, Vol. ED-38(3), pp.603-610, 1991.

14. C. Hu, "Lucky-electron model of channel hot electron emission," in *IEDM Tech. Dig.*, pp.223-226, 1979.

15. B. Eitan, and D. Frohman-Bentchkowsky, "Hot-Electron injection into the oxide in n-channel MOS devices," *IEEE Trans. Electron Dev.*, Vol. ED-28(3), pp.328-340, 1981.

16. P. E. Cottrell, R. R. Trooutman, and T. H. Ning, "Hot-electron emission in n-channel IGFET's," *IEEE Trans. Electron Devices*, vol. 26, no. 4, pp. 520-532, 1979.

17. G. A. Baraff, "Distribution functions and ionization rates for hot-electrons in semiconductors," *Phys. Rev.*, vol. 128, no. 6, pp. 2507-1517, 1962.

18. S. Tam, P. K. Ko, C. Hu, and R. Muller, "Correlation between substrate and gate currents in MOSFET's," *IEEE Trans. Electron Devices*, vol. 29, no. 11, pp. 1744-1744, 1982.

19. E. Takeda, H. Kune, T. Toyabe, and S. Asai, "Submicrometer MOSFET structure for minimizing hot-carrier generation," *IEEE Trans. Electron Devices*, vol. 29, no. 4, pp. 611-618, 1982.

20. K. Hess and C. T. Sah," Hot carriers in Silicon surface inversion layers," *J. Appl. Phys.*, vol. 45, p. 1254, 1974.

21. K. R. Hofmann, C. Werner, W. Weber, and G. Dorda,"Hot-electrons and hole-emission effects in short n-channel MOSFET's," *IEEE Trans. Electron Devices*, vol. 32, no. 3, pp. 691-699, 1985.

22. B. Riccò, G. Torelli, M. Lanzoni, A. Manstretta, H. E. Maes, D. Montanari, and A. Modelli, "Nonvolatile multilevel memories for digital applications," *Proc. of the IEEE*, Vol. 86, N. 12, pp.2399-2421, 1998.

23. L. Esaki, "Long journey into tunneling," in *Proc. IEEE*, vol. 62, pp. 825-831, 1974.

24. J. Moll, *Physics of Semiconductors*. New York: McGraw-Hill, 1964.

25. M. Lenziger and E.H. Snow, "Fowler-Nordheim tunneling into thermally grown SiO_2," *J. Appl. Phys.*, vol. 40, no. 1, pp. 278-283, 1969.

26. P. Olivo, T. Nguyen, and B. Riccò, "High-field-induced degradation in ultra thin SiO_2 films," *IEEE Trans. Electron Dev.*, Vol. ED-35(12), pp. 2259-2267, 1988.

27. J. Tang and K. Hess, "Theory of hot electron emission from silicon into silicon dioxide," *J. Appl. Phys.*, vol. 54, pp. 5145-5151, 1983.

28. J. Suñè, P. Olivo, and B. Riccò, "Quantum-mechanical modeling of accumulation layers in MOS structure," *IEEE Trans. Electron Devices*, vol. 39, no. 7, pp. 1732-1738, 1992.

29. P. Cappelletti, R. Bez, D. Cantarelli, and L. Fratin, "Failure mechanisms of Flash cell in program/erase cycling," *IEDM Tech. Dig.*, pp. 291–294, 1994.

30. K. T. San, C. Kaya, D. K. Y. Liu, T. P. Ma, and P. Shah, "A new technique for determining the capacitive coupling coefficients in FLASH EPROM's," *IEEE Electron Device Lett.*, vol. 13, no. 6, pp. 328–331, 1992.

31. S. Aritome, R. Shirota, G. Hemnik, T. Endoh, and F. Masuoka, "Reliability issues of Flash memory cells," in *Proc. IEEE*, vol. 81, pp. 776–788, May 1993.

32. C. Papadas, G. Ghibaudo, G. Pananakakis, C. Riva, P. Ghezzi, C. Gounelle, and P. Mortini, "Retention characteristics of single-poly EEPROM cells," in *Proc. European Symp. Reliability of Electronic Devices, Failure Physics and Analysis (ESREF)*, Bordeaux, France, Oct. 7–10, 1991, pp. 517–522.

33. A. Watts, "Built-in reliability for 10 FITS performance on EPROM and Flash memory," SGS-Thomson Microelectronic, Agrate Brianza, Italy, Tech. Art. TA 109, Nov. 1991.

34. P. L. Hefley and J. W. McPherson, in *Proc. IRPS*, 1988, p. 176.

35. G. Crisenza, R. Annunziata, E. Camerlenghi, and P. Cappelletti, "Non volatile memories: Issues, challenges and trends for the 2000's scenario," in *Proc. ESSDERC'96*, G. Baccarani and M. Rudan, Eds. Bologna, Italy: Editions Frontieres, 1996, pp. 121–130.

36. S. Mori, Y. Yamaguchi, M. Sato, H. Meguro, H. Tsunoda, E. Kamiya, K. Yoshikawa, N. Arai, and E. Sakagami, "Thickness scaling limitation factors of ONO interpoly dielectric for nonvolatile memory devices," *IEEE Trans. Electron Devices*, vol. 43, no. 1, pp. 47–53, 1996.

37. S. Maramatsu, T. Kubota, N. Nishio, H. Shirai, M. Matsuo, N. Kodama, M. Horikawa, S. Saito, K. Arai, and T. Okazawa, "The solution of over-erase problem controlling poly-Si grain size modified principles for Flash memories," *IEDM Tech. Dig.*, 1994, pp. 847–850.

Chapter 3

DC CONDITIONS: READ
Theory, compact modeling and circuit simulation

The main purpose of this chapter is to introduce the basic principles and the main guidelines to have an effective compact modeling (CM) of the Floating Gate (FG) memory devices in DC conditions. In particular, according to the purposes of this book, the real possibility to implement the CMs of FG devices into Spice-like circuit simulators will be accurately investigated, being circuit simulators a useful tool to gain a basic and wide understanding of the functioning and operations of electronic devices. In fact, Spice-like circuit simulators can be frequently and easily used without excessive computation time either to evaluate the single devices' behavior or to simulate more complex circuits, thus constituting an interesting tradeoff among ease of use, computational effort and simulation accuracy.

The key issue for a correct modeling of the electrical operations of FG devices is given by the accurate calculation of the FG voltage. In fact, once the FG voltage is known, the modeling of a FG device reduces to the modeling of an MOS transistor, whose gate voltage is equal to the FG one. Since the FG is completely surrounded by an insulator layer - so that it cannot be accessed- , the FG voltage cannot be fixed and its value, which has to be calculated, is determined by the capacitive coupling of the FG electrode with the other electrodes of the devices (Drain, Body, Source, and Control Gate).

The chapter is organized as follows. In Paragraph 3.1, the classical approach reported in the literature to calculate the FG voltage will be presented and its limits accurately reviewed and discussed. Then, a new approach to perform this task, which is more suitable for circuit simulation implementations, will be proposed (Paragraph 3.2). Simulation results proving the accuracy of this model will be presented, and the main advantages of this new modeling strategy will be discussed and compared with the most common approaches (Paragraph 3.3).

1. TRADITIONAL FG DEVICE MODELS

The analysis and the theoretical modeling of FG memory cells are well-known subjects in the literature [1-2]. Particularly, many efforts have been spent in the past to model the program and erase transients of the FG devices [2-5]. On the contrary, very few works have been proposed in the literature to address the task of simulating the DC behavior of the FG memory cells [6-7].

In this scenario, the pioneering work of Kolodny et al. [2] is certainly the most important one, as it outlines the basic theory of the classic FG voltage calculation method. The main details of this method will be summarized in Paragraph.1.1, whereas the following Paragraph 1.2 will explain how to extend to FG devices the classic formula of the drain-source current developed for MOS transistors. Paragraph 1.3 will discuss the main limits of the standard FG voltage calculation procedure, clearly demonstrating that for an accurate modeling of the FG devices some other methods to estimate the FG voltage are required.

1.1   The classical FG voltage calculation method

The calculation of the FG voltage is traditionally tackled adopting the schematic cross-section shown in Fig. 3-1.

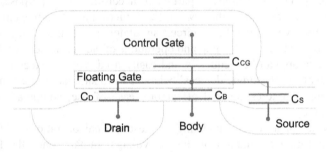

Figure 3-1. Schematic cross section of a EEPROM memory cell, where C_D, C_S, C_B, C_{CG} are the capacitances between FG and D, S, B and CG respectively.

The upper gate is the Control Gate (CG) and the lower gate, completely isolated within the gate dielectric, is the Floating Gate (FG), which acts as a potential well. If a charge is forced into the well, it cannot move from there without applying any external force: the FG stores charge [1]. The simple model shown in Fig. 3-1 (known also as capacitive coupling coefficient model) helps in understanding the electrical behavior of a FG device. C_{CG}, C_S, C_D, and C_B are the capacitances between FG and CG, Source (S), Drain (D) and Body (B), respectively. If no charge is stored in the FG, i.e. $Q = 0$:

$$Q = 0 = C_{FG}(V_{FG} - V_{CG}) + C_S(V_{FG} - V_S) + C_D(V_{FG} - V_D) + + C_B(V_{FG} - V_B) \,(1)$$

where V_{FG} is the potential on the floating gate, V_{CG} is the potential on the control gate, V_S, V_D, V_B are potentials on S, D and B, respectively. If we name $C_T = C_{FG} + C_D + C_S + C_B$ the total capacitance of the FG, and we define $\alpha_J = C_J / C_T$ as the coupling coefficient relative to the electrode J, where J can be one among CG, D, S, and B, the FG potential is given by

$$V_{FG} = \alpha_{CG}V_{CG} + \alpha_D V_D + \alpha_S V_S + \alpha_B V_B . \qquad (2)$$

It is interesting to note that the Floating Gate voltage does not depend only on the control gate voltage, but also on source, drain and bulk potentials. Moreover, if source and body are both grounded (3.2) can be rearranged and reduces to

$$V_{FG} = \alpha_{CG}\left(V_{CG} + f \cdot V_{DS}\right), \qquad \text{where} \qquad f = \frac{\alpha_D}{\alpha_{CG}} = \frac{C_D}{C_{CG}}. \qquad (3)$$

Several procedures have been proposed in the literature to derive the capacitive coupling coefficients that cannot be directly measured being the Floating Gate electrically isolated. Such extraction procedures are often cumbersome and inaccurate, and they will be discussed in the next Paragraph 1.3.

1.2   Drain current calculation

Device equations for the FG MOS transistor can be obtained from the conventional MOS transistor equations simply by replacing MOS gate voltage, V_{GS}, with V_{FG}, and transforming the device parameters, such as threshold voltage, V_T, and conductivity factor, β, to values measured with respect to the control gate: $V_T^{FG} = \alpha_{CG} \cdot V_T^{CG}$ and $\beta^{FG} = \beta^{CG}/\alpha_{CG}$ [1].

In this way, the current-voltage (I-V) equations of FG MOS transistor in both Triode Region (TR) (4) and Saturation Region (SR) (5) can be easily derived from the ones of a conventional nMOS transistor [8].

$$I_{DS} = \beta^{CG}\left[\left(V_{CG} - V_T^{CG}\right)V_{DS} - \left(f - \frac{1}{2\alpha_{CG}}\right)V_{DS}^2\right] \quad V_{DS} < \alpha_{CG}\left(V_{CG} + fV_{DS} - V_T^{CG}\right) \quad (4)$$

$$I_{DS} = \frac{\beta^{CG}}{2}\alpha_{CG}\left(V_{CG} + fV_{DS} - V_T^{CG}\right)^2 \qquad V_{DS} \geq \alpha_{CG}\left(V_{CG} + fV_{DS} - V_T^{CG}\right) \quad (5)$$

Comparing these equations to those of a conventional MOS transistor, some differences, mainly due to the capacitive coupling between drain and floating gate, can be observed [8].

1. The FG MOS transistor can conduct a relevant current even when $V_{CG}-V_S<V_T$, because the channel can be turned on by the drain voltage through the $f \cdot V_{DS}$ term in (3). This effect is usually referred as "drain turn-on".

2. The saturation region for a conventional MOS transistor is where I_{DS} is essentially independent of V_{DS}. This is no longer true for the FG transistor, in which I_{DS} continues to rise as the drain voltage increases. In other words, this means that no saturation occurs, and I_{DS} depends on V_{DS} also in saturation conditions. Note that for the FG transistor the boundary between triode and saturation regions is expressed by $V_{DS}=\alpha_{CG}(V_{CG}+f \cdot V_{DS}-V_T)$, compared to the conventional nMOS condition: $V_{DS}=V_{GS}-V_T$.

3. In saturation region, the transconductance $g_m = \alpha_{CG} \cdot \beta^{CG}(V_{CG} + f \cdot V_{DS} - V_T)$ increases with V_{DS}, in contrast to conventional MOS transistors, where g_m is relatively independent of V_{DS}.

If some charge is stored in the FG, i.e. $Q \neq 0$, all the hypotheses made above hold true, and the following modifications need to be included to evaluate V_{FG} and V_T^{CG}.

$$V_{FG} = \alpha_{CG} V_{CG} + \alpha_D V_D + \alpha_S V_S + \alpha_B V_B + \frac{Q}{C_T} \tag{6}$$

$$V_T^{CG} = V_{T0}^{CG} - \frac{Q}{C_{CG}} \tag{7}$$

V_{T0} is the threshold voltage when Q=0, and equation (7) showing the linear dependence of the threshold voltage on the charge injected into the FG, can be rewritten in terms of the threshold voltage shift, $\Delta V_T = V_T - V_{T0} = -Q/C_{CG}$.

Finally, by inserting (7) into (4) and (5), the current-voltage equations of FG MOS transistor in TR and SR can be easily modified to account for $Q \neq 0$.

$$I_{DS} = \beta^{CG}\left[\left(V_{CG} - V_{T0}^{CG} + \frac{Q}{C_{CG}}\right)V_{DS} - \left(f - \frac{1}{2\alpha_{CG}}\right)V_{DS}^2\right]$$
$$V_{DS} < \alpha_{CG}\left(V_{CG} + fV_{DS} - V_{T0}^{CG} + \frac{Q}{C_{CG}}\right) \tag{8}$$

$$I_{DS} = \frac{\beta^{CG}}{2}\alpha_{CG}\left(V_{CG} + fV_{DS} - V_{T0}^{CG} + \frac{Q}{C_{CG}}\right)^2$$
$$V_{DS} \geq \alpha_{CG}\left(V_{CG} + fV_{DS} - V_{T0}^{CG} + \frac{Q}{C_{CG}}\right) \tag{9}$$

Equations (8)-(9) show that the role of the charge injected into the FG is to shift the I-V curves of the memory cell by the quantity $-\Delta V_T = Q/C_{CG}$. As explained in Chapter 2, this effects allows to read the information stored in a FG memory cell simply biasing it with a constant voltage (usually $V_{CG} \sim 4\text{-}5V$, $V_{DS} \sim 1V$) and comparing the current driven by the memory cell to that of a reference cell.

1.3 Limits of the capacitive coupling coefficient method

Although the classic FG voltage calculation method explained in Paragraph 1.1 is widely used, the accuracy limits of Equation (1) are not widely known and deserve to be underlined. Schematically, the reasons of the poor accuracy of the standard FG voltage calculation method are mainly two. Firstly, since the FG is electrically isolated and cannot be directly accessed, the capacitive coupling coefficients are not so easy to evaluate, and the methods proposed for their extraction are often cumbersome and inaccurate [2,9-14]. Then, the capacitive coupling ratios, that are defined as constants $\alpha_J = C_J/C_{TOT}$, depend on the bias, and therefore, neglecting their bias-dependence can lead to significant errors in the FG voltage calculation [9,10,11,15].

1.3.1 The capacitive coupling coefficient extraction procedure

Since the FG is electrically isolated and cannot be directly accessed, several calculation procedures have been established to evaluate the capacitive coupling coefficients of FG memory cells. Indeed, these methods are mainly focused on the estimate of the α_{CG}, as this coefficient is that of major interest for its impact on the modeling of FG devices [2,9,11,13,14]. On the other hand, few calculation procedures have been proposed to estimate the source α_S and drain α_D coupling coefficients, and in all the cases they require the knowledge of the previous calculated α_{CG} [12-13]. Finally, the α_B determination is still an open issue.

1.3.1.1 The control gate coupling coefficient

Several methods have been proposed in the literature to estimate the CG coupling coefficient. The most used ones calculate α_{CG} from the ratios between *threshold voltages*, *gains*, *transconductances* or *subthreshold slopes* measured on the dummy cell and FG memory cell [2,13]. Since such methods rely on the perfect matching of the dummy and FG cell, they suffer from the small variations between the two devices that introduce errors in the α_{CG} estimate becoming more severe with the scaling of devices [2,11-15]. Nevertheless, in the industry these techniques (which can be implemented in automatic routines) are largely preferred for their simplicity over the complexity of the methods using only the FG device, which are generally more complex. In fact, these last ones require either a great number of measurements or complex procedures of data analysis: in [12], for example, the coupling coefficients are calculated from the erase characteristics and junction leakage current measurements; in [11], CG and D ratios are determined from the solution of a linear system of two equations coming from drain turn-on measurement and programming characteristics of the cell; the method proposed in [14] is based on the program pulse dependence on the hold time, thus appearing rather complex.

For this reason, although the differences between the dummy and the FG cells can lead to significant errors in the CG coefficient estimate, the methods using the dummy cell are largely preferred for their simplicity. Further, besides the major limitations discussed above, these methods have also other specific application issues: the *threshold voltage* and *gain* methods are strictly affected by the non-zero value of FG charge [2,13], and the *transconductance* one is very sensitive to short-channel effects and mobility degradation [2], so that errors introduced cannot be neglected in advanced devices [13]. To conclude, the *subthreshold* method seems to be the most accurate, even though it can be used only when the device is in subthreshold conditions and the error affecting the α_{CG} estimate is not negligible [15].

1.3.1.2   Drain and source capacitive coefficients

There are only two methods known by the authors to evaluate D and S coupling coefficients [9,12,13]. The first method reported in [12] is based on the fact that the tunnel voltage between FG and S is independent on the initial conditions (V_S and initial charge in FG), and establishes a relation among the S capacitive coupling ratio, the difference between erase voltages, ΔV_S, and the difference between corresponding threshold voltages after erase, $\Delta(\Delta V_T)$.

$$\frac{1-\alpha_S}{\alpha_{CG}} = \frac{\Delta(\Delta V_T)}{\Delta V_S} \tag{10}$$

The second method takes advantage of the fact that the drain junction leakage current is controlled by $V_{FG}-V_D$ and V_D-V_B [9,12-13]. By keeping constant V_D and V_B, the S bias has to be adjusted (ΔV_S) to keep the drain current constant for each change in the CG bias, ΔV_{CG}. Thus, a simple law to evaluate both α_S and α_D can be extracted from (1):

$$\frac{\alpha_{S(D)}}{\alpha_{CG}} = \frac{\Delta V_{CG}}{\Delta V_{S(D)}} \tag{11}$$

Unfortunately, both methods are indirect and require the knowledge of α_{CG}, thus suffering from the same errors affecting the FG coupling coefficient estimate. Besides, there are other issues related to the real application of these techniques: the first method cannot be applied successfully on Flash memories whose threshold voltage shift depends negligibly on V_S, whereas the second one is much more sensitive to ΔV_{CG}. To conclude, it is worth reminding that α_S and α_D estimated by these methods are affected by errors that are larger than those affecting the α_{CG} estimate, resulting too inaccurate to provide physically reasonable results [15].

1.3.2   The bias dependence of the capacitive coupling coefficients

As reported in the literature, although the capacitive coupling ratios are defined as constants $\alpha_J = C_J/C_{TOT}$, they depend significantly on the CG, D, S, and B biases applied to the FG device [9,10,11,15]. For this reason, considering them as constant determines a significant error in the FG voltage estimate.

The analysis of the bias dependence of α_{CG} reported in the literature has shown that this coupling coefficient depends on CG and B voltages (it shows an abrupt decrease moving from subthreshold to above-threshold conditions), whereas it does not depend significantly on V_S and V_D [15]. Schematically, α_{CG} has two different values in sub-threshold and above-threshold conditions, whose difference (5-10%) can originate a large error in the V_{FG} estimate. For example, if α_{CG} is evaluated by means of the *subthreshold slope* method and then it is used to calculate V_{FG} during program operations (above-threshold conditions), the estimated FG voltage can be higher than the real one by more than 0.5V, thus seriously compromising the accuracy of the V_{FG} estimate [2,9,13]. In this respect, particular care has to be taken using Equation (1) to calculate V_{FG} in read conditions, since the α_{CG} value which has to be considered depends on the programmed or erased state of the FG memory cell, which might be in sub-threshold regime, if programmed, or above-threshold regime, if erased.

From the physical stand point, the α_{CG} bias dependence is related to the V_{CG} - V_B dependence of the FG to B capacitance (C_B). In fact, when $V_{CG}<V_T$, the substrate depletion capacitance (C_{SUB}) becomes comparable to the oxide one (C_{OX}), thus reducing significantly C_B, which is given by the series of C_{SUB} and C_{OX}. On the contrary, in inversion C_{SUB} is very large, and therefore C_B is approximately given by the C_{OX} alone. Since the other capacitances (C_D, C_S and C_{CG}) are almost constant with respect to CG and B biases, the abrupt decrease of $\alpha_{CG} = C_{CG}/C_T$ moving from sub-threshold to above-threshold conditions is due the increase of C_B, i.e. C_T.

Since the α_{CG} decrease is related only to the increase of C_B and C_T, the same bias dependence on V_{CG} and V_B shown by α_{CG} is expected also for α_D and α_S. Besides the abrupt decrease of α_S and α_D passing from sub-threshold to above threshold conditions, the D and S coupling coefficient has shown also a strong increase at high V_{CG} and low V_S and V_D. From the physical point of view, such strong increase is due the growth of C_D and C_S beyond their *geometrical* value, which can be calculated summing the overlap and fringing capacitance contributions. The increase of C_D and C_S is related to the uniform electron layer established along the whole oxide interface in these particular bias conditions (high V_{CG} and low V_S and V_D), that extend the regions contributing to the C_D and C_S to the adjacent channel area, practically increasing the effective overlap region beyond the metallurgical junction [15].

Similarly to other coupling coefficients, also α_B changes values going from sub-threshold to above-threshold regions, dropping to zero when $V_{CG}-V_S>V_T$. The physical reasons of the abrupt drop of α_B are related to the channel potential at the oxide interface that, when the channel is strongly inverted, is independent on the body voltage, being fixed uniquely by V_S and V_D. Thus, since any V_B variation does not change the voltage drop across the tunnel oxide, i.e. V_{FG}, the body coefficient, which is the measure of the V_B influence on the FG voltage, falls to zero in these bias conditions. Moreover, differently from other coupling coefficients that do not show significant differences in accumulation condition (erase operations), α_B doubles its value compared to the one derived in subthreshold conditions. Again, this is due to the $V_{CG} - V_B$ bias dependence of C_B. In fact, since in accumulation C_B is practically equal to C_{OX} being the substrate capacitance C_{SUB} very large, the body coefficient, which is defined as C_B/C_T, doubles [15].

To conclude, the use of the coupling capacitive methods, and specifically of Equation (2) introduces in the V_{FG} calculations errors that are as larger as the bias conditions are more different from those at which coupling coefficient have been extracted [9-11,15].

2. THE CHARGE BALANCE MODEL

As known, the effectiveness of FG device modeling is directly correlated to the accuracy of the FG voltage calculation. However, as highlighted in the previous Paragraph, the calculation procedure classically adopted to evaluate the FG voltage, i.e. the capacitive coupling coefficient method, shows significant limits compromising the accuracy of the FG potential calculation, and therefore of the whole FG device modeling.

In this respect, FG memory models proposed in the literature in the past years do not overcome this strict limit. To this regard, a first example is given by the model of drain-source current presented in [6]. This model neglects completely the short-channel effects, the mobility reduction due to the normal field, and the velocity overshoot, so that its excessive simplifications lead to large errors increasing with the scaling down of the technology, thus representing a serious limit for the application of the model on future generations. Moreover, this model cannot be simply implemented in Spice-like circuit simulators, and this poses serious difficulties to its use in the industrial environment. Although the above limits are not shown by the CM proposed in [7], this model employs the coupling coefficient method to estimate the FG voltage, and therefore its simulation results suffer of the accuracy limits highlighted in Paragraph 1.3. Moreover, this model refers to the specific case of a Flash cell and, therefore, it cannot be adopted to model an EEPROM cell. In this scenario, since no easy-to-use and accurate CM of FG memory cells was available, to simulate circuits including FG devices in the industry it was a common practice to replace the FG memory cells with simple MOS transistors, whose threshold voltage was manually modified for reproducing the programmed or erased state of the memory cell.

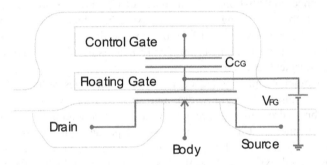

Figure 3-2. New model of the FG memory cell, comprised of the FG-CG capacitor, C_{CG}, the MOS transistor equivalent to the dummy cell, and the voltage controlled voltage source, V_{FG}, connected between the FG and ground.

Only recently, to respond to the growing demand of FG device CMs and also to estimate more accurately the FG potential which is fundamental for the correct modeling of FG devices, a new CM of FG memories based on a new FG voltage calculation procedure has been proposed in the literature [16]. This model, which is suited for circuit simulator implementations, is comprised of three elements (see Fig. 3.2): an MOS transistor whose Source, Body, Drain are Source, Body, Drain, of the cell, and Gate is the FG of the cell; a capacitor connected between FG and Control Gate of the cell; a voltage-controlled voltage source, V_{FG}, between ground and FG, which is necessary to overcome the problem of simulating a capacitive net in DC conditions. In fact, Spice-like circuit simulators are not capable to solve the net with a simple series of two capacitors in DC conditions. In other words, the FG node has to be biased to its correct value by an external source: the voltage-controlled voltage source, V_{FG}, which constitutes the core of the model in DC conditions. As it will be explained in the next Paragraph 2.1, this circuit element implements the new FG voltage calculation procedure, which does not use the fixed capacitive coupling coefficients, thus improving the FG voltage calculation. The main advantages of this new modeling strategy will be discussed in comparison to the classical capacitive coupling coefficient approach in the Paragraph 2.2, while simulation results will be compared to experimental curves in the Paragraph 2.3.

2.1   The Floating Gate voltage calculation procedure

A new method taking advantage from the implementations of the model as a simple Spice-like sub-circuit has been developed in [16] to calculate the FG voltage. The new FG voltage calculation procedure is based on the solution of the charge balance equation at the floating gate node. That is, the charge on the MOS gate, Q_G, plus the charge on the bottom plate of the FG-CG capacitor, is equal to the charge forced in/out the Floating Gate during program/erase operations, Q_{FG}, which in DC conditions is constant and depends on the state of the memory cell.

$$Q_G\left(V_{FG}, V_S, V_D, V_B\right) + C_{CG}\left(V_{FG} - V_{CG}\right) = Q_{FG} \tag{12}$$

The charge on the gate of the MOS transistor, Q_G, is a complex function of S, D, B and FG voltages (V_S, V_D, V_B and V_{FG}, respectively). It can be evaluated using the charge equations of the compact MOS transistor model adopted to simulate the dummy cell. To this regard, several MOS CMs can be used (Philips MM9/11 [17-18], BSIM3v3/4 [19-20], EKV [21-22], SP [23-24], HiSIM [25]), and the choice of the right model involves different factors: the knowledge and the practice to use a specific MOS models, particularly regarding to the parameter extraction procedure; the tradeoff between accuracy and simplicity (number of parameters); the computational complexity.

In some circuit simulators, the MOS transistor charges at S, D, B and G electrodes are directly available, so that their calculation is not required, and the FG voltage calculation procedure is strongly simplified. The solution of the charge balance equation corresponds to the finding of the zero of the following function of V_{FG}, which has been obtained from (11) rearranging its terms.

$$F(V_{FG}) = Q_G\left(V_{FG}\right) + C_{CG}\left(V_{FG} - V_{CG}\right) - Q_{FG} \tag{12}$$

$F(V_{FG})$ is always monotonic versus V_{FG} for all the combinations of V_S, V_D, V_B and V_{CG} of interest. This could also be deduced looking at the terms constituting $F(V_{FG})$, which are monotonically increasing functions of V_{FG}, expect Q_{FG}. The monotonic trend of $F(V_{FG})$, which assures that the charge balance equation has a unique solution (otherwise the physical meaning would be lost), has been usefully exploited to develop the numeric algorithm to solve (12), i.e. to calculate V_{FG}. For details about the V_{FG} calculation algorithm the reader can refer to [16]. To conclude, it is worth recalling again that this procedure calculates the FG voltage more accurately compared to previous models relying on the constant capacitive coupling coefficient method, thus significantly improving the whole modeling of the FG device.

2.2   Advantages and scalability

The new approach of the model proposed in [16], and particularly its new V_{FG} calculation procedure, gives further advantages compared to standard models relying on the capacitive coupling coefficient method.

1. This model is easily scalable, since scaling rules are already included in the compact MOS model adopted and they do not affect directly the V_{FG} calculation routine.
2. This model is very simple to implement, since it uses standard circuit elements whose parameters can be determined by applying the MOS parameter extraction procedure to the dummy cell (a FG memory cell whose FG and CG are short-circuited), and the few other additional parameters can be easily estimated from cell layout and cross section.
3. The accuracy of the model depends mainly on the compact MOS model adopted: in this respect, the model takes advantage of the many efforts to improve and scale MOS CMs.
4. The computation time is comparable to that of a simple MOS transistor, so that it can be used conveniently and efficiently in Spice-like circuit simulators.
5. This model can be easily extended to simulate transient behaviors of FG memories by adding a suitable set of voltage controlled current sources to its basic structure, provided that reliable models of program/erase mechanisms are given (see Chapter 4).
6. In the same way, this model can be used to evaluate the impact of the leakage current due to electrical (SILC) and radiation (RILC) degradation of the tunnel oxide on the memory cell reliability, provided that reliable models of RILC and SILC are available.

2.3   Parameter extraction

The procedure to extract the parameters of device CMs is not a "push-button" task. For Floating Gate devices, this task is even more complex than for standard MOS transistors. In the following, we will shortly describe the suggested procedure for the FG device parameter extraction, which for clarity has been schematically divided into three parts.

The first part deals with the extraction of the DC parameters of the MOS transistor modeling the *dummy cell*, which is the cell where FG and CG are short-circuited. A set of dummy cells having different widths and lengths close to the typical cell's width and length must be designed in the test-pattern designed for parameters extraction. The usual MOS extraction strategy can be used with reasonable results paying attention to the slightly different physics of the *dummy cell* compared to a standard MOS transistor. In fact, the narrow and short geometry, the lack of LDD and Pocket Implant determine a less ideal behavior such as larger DIBL effect and higher multiplication current.

Moreover, particular care has to be devoted to extract the overlap capacitance values. These parameters can be evaluated from Capacitance-Voltage (CV) measurements performed on a large enough array of dummy cells. In order to separate the Drain (D) and Source (S) overlap capacitance contributions from the channel one, it is necessary to perform the CV measurement in accumulation conditions separating the source and drain current from the Bulk (B) one. From the accumulation branch of such CV measurement it is possible to calculate the D-FG overlap capacitance. It is important to pay attention to the parasitic interconnect capacitance contribution that could affect the measurement. In this case it would be better to use a dedicated layout to minimize such contribution. By repeating the measurement for the source terminal of the array, also the S-FG overlap capacitance can be estimated. The B-FG capacitance can be evaluated as well by analyzing in this case the inversion branch of the CV curve. Note that as overlap capacitances are very small, their evaluation is particularly critical and needs particular attention. Numerical simulation tools can help this difficult task. In fact, if the measurement is difficult as we said, the capacitance simulation is quite easy and accurate, once the structure to be simulated is well described. The weak point of this approach is that the simulations have still to be carried out in 2 dimensions, since there is no mature 3-D simulation tool available.

The second part of the procedure deals with the extraction of additional parameters, that except the CG-FG capacitance (C_{CG}) depend on the kind of FG memory considered. For example, if we consider an EEPROM memory cell, the additional parameters that need to be estimated are: 1) the area of the tunneling region; 2) the tunnel oxide thickness; 3) the doping levels of the drain well and the FG. For converse, the additional parameters of a Flash memory cell are: 1) the areas of S-FG, D-FG, and channel-FG overlap regions; 2) the doping levels of S and D wells, channel and FG. Generally, these parameters are either directly evaluated from the layout of the cell (C_{CG}, tunnel and overlap region areas), or straightly derived from the process recipe (doping). Unfortunately, there is no dedicated procedure that can be utilized to extract such parameters from measurements performed directly on FG devices. Sometimes, dedicated measurements performed on MOS capacitors designed ad-hoc to reproduce specific portions of the FG devices (for example, the tunnel capacitance in EEPROM memory cell) can be used to extract and verify parameter values.

Finally, the third and final step of the parameter extraction procedure consists in the verification of the extracted parameter values. To this purposes, the MOS parameters of the dummy cell extracted in the first step of the procedure are checked by comparing simulation results to measurements performed on *dummy cells*. Then, measurements performed on actual FG devices are used to verify the estimated additional parameters. Particularly, I_{GS}-V_{DS} and I_{DS}-V_{DS} curves are employed to verify and (if it is the case) achieve a finer calibration of C_{CG}, adopting as a first "fitting parameter" the charge in the FG, which might be the residual charge due the device manufacturing process. Then, the other additional parameters (tunnel oxide thickness, tunneling region area, S-FG, D-FG, and B-FG overlap area) can be verified by comparing simulations and measurements of program/erase characteristics of the specific FG memory cells considered.

3. SIMULATION RESULTS

Besides the several advantages highlighted in the previous Paragraph, the charge balance model allows also a very accurate simulation of the DC characteristics of EEPROM and Flash memory cells. EEPROM memory cells considered to test the simulation capability of the model have been manufactured in 0.35 μm technology with the following characteristics: gate and tunnel oxide thicknesses are 20 and 7 nm, respectively; interpoly dielectric capacitance is $C_{GC}=3fF$; width and length of the cell are

W=0.3µm and L=0.75µm, respectively. Three different kinds of Flash memories manufactured in 0.25 and 0.18 µm technology have been also selected to test the simulation capability of the model. Their main characteristics are: the gate oxide thickness is around 10 nm; the interpoly capacitance spans from ~0.8 to ~0.4fF; L and W vary between 0.425 µm and 0.3 µm, and 0.3 µm and 0.16 µm, respectively. Note that the smallest Flash memory devices considered (very short and narrow conductive channel) represent a severe test to evaluate the model capability to simulate future FG memory generation.

Examples of simulation capabilities of the charge balance model are shown compared to experimental data in Figures 3-3(a)-(d). Simulations have been obtained by assuming a small fixed charge in the Floating Gate (Q_{FG}=-0.65 fC), which is probably due to residual charge after P/E cycling. The compact MOS models used for these examples is Philips MM9. As shown in these figures, the agreement between I-V measurements and simulations is excellent without the need of any free parameter to adjust fitting quality.

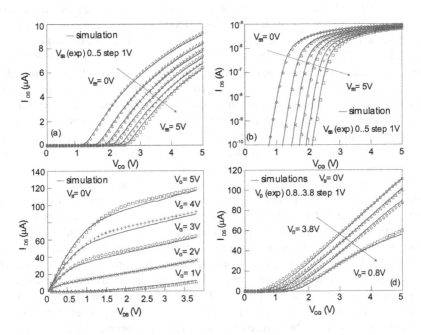

Figure 3-3. Experimental curves (symbols) and model simulations (solid lines) obtained assuming Q_{FG}=-0.65fC as residual charge in the FG are shown for an 0.35µm EEPROM memory cell (W=0.3µm, L=0.75µm, C_{CG}=3fF).

Two different approaches have been followed to model the EEPROM memory cell. As the EEPROM memory cell is constituted by the FG transistor in series with the select transistors, this last one has also to be taken into account to simulate correctly the behavior of the EEPROM memory cell. This can be done in two different ways.
1. The select transistor is directly included into the schematic, and the circuit model of the memory cell is thus constituted by the series of select and FG transistors. Obviously, in this case two different model-cards have to be extracted: one for the

select transistor, using some test structure to emulate it, and one for the FG transistor, using the *dummy* cell. For this reason, adopting this approach the implementation-computation time can slightly increase compared to the case the alone FG transistor.

2. Since the gate of the select transistor is highly biased (~10-15 V) when the memory cell is addressed, the select transistor can be simply modeled by a small resistance connected in series with the drain of the FG transistor. Therefore, the contribution of the select transistor can be effectively accounted for by a slight increase of the drain resistance of the transistor modeling the *dummy* cell. Conveniently, this can be done automatically by applying the MOS parameter extraction procedure to the series of the dummy cell and the select transistor highly biased (~10-15 V).

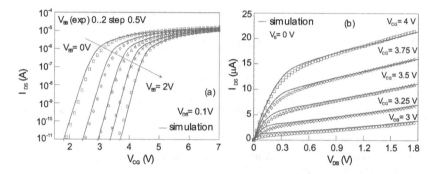

Figure 3-4. Experimental curves (symbols) and model simulations (solid lines) obtained assuming Q_{FG}=-0.13fC as residual charge in the FG are shown for a 0.25μm Flash memory cell (W=0.25μm, L=0.375μm, C_{CG}=0.8fF).

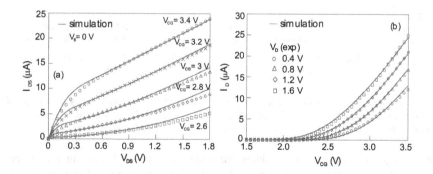

Figure 3-5. Model simulations (solid lines) and experimental curves (symbols) measured on a 0.18μm Flash memory cell (W=0.22μm, L=0.3μm, C_{CG}=0.56fF).

The results obtained in the two cases are very similar (data in Figures 3-3(a)-(d) refer to the latter case), and therefore, the second approach is preferable as it requires only the extraction of one model-card (*dummy* cell).

This problem does not occur when modeling Flash memory cells, since the standard Flash cell is composed by a single FG transistor, and therefore the extraction procedure can be easily applied to the *dummy* cell alone. Again, as shown in Figures 3-4, 3-5, 3-6, the agreement between simulations and measurements is excellent in every possible bias combination, above and sub-threshold, with and without substrate bias, and for each of the three Flash memory cell considered. As for EEPROM, a small residual charge, probably due to P/E cycling of samples, has been assumed in the FG (Q_{FG}=-0.13 fC) for one of the Flash memory cells considered. The compact MOS models adopted is Philips MM9. No free parameters have been adopted to improve the agreement between measurements and simulations.

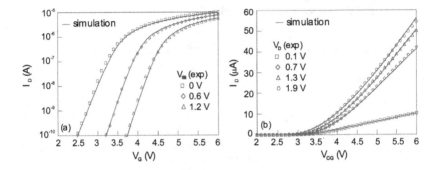

Figure 3-6. Model simulations (solid lines) and experimental curves (symbols) measured on a 0.18µm Flash memory cell (W=0.16µm, L=0.3µm, C_{CG}=0.43fF).

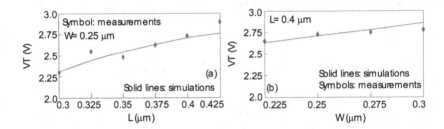

Figure 3-7. Scaling trends of threshold voltage, V_T, measured (symbols) and simulated by the charge balance model (lines) by keeping constant the FG charge density on a 0.25µm Flash memory cell, whose length L and width W have been scaled from 0.425µm to 0.3µm, and from 0.3µm to 0.225µm, respectively.

The charge balance model simulates correctly also W and L scaling effects on FG memory cells. To this regard, Figures 3-7(a)-(b) show the trends of the measured and simulated threshold voltage, V_T, which is defined as the CG voltage that allows I_{DS}=1 µA/µm channel width, when V_{DS}=1 V. Simulations have been performed by keeping constant the FG charge density, since we have assumed that this quantity is related to the damage induced by the manufacturing process of the memory cell. As reported in Figures 3-7(a)-(b), experimental V_T curves do not show a clear trend versus W and L. On the contrary, V_T simulations rise monotonically on increasing W and L, as expected from basic theory. We have found that such disagreement is due to a slightly different amount of the FG charge in the different samples we have measured. In fact, with a good fitting of the value of Q_{FG}, the discrepancy between simulations and measurements disappears. On the other hand, such differences in the FG charge are largely expected, and they are attributed to the random nature of the damage process, as well as to the different "history" of each sample.

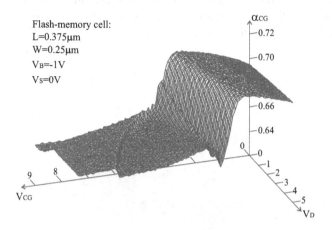

Figure 3-8. CG coupling coefficient of a 0.25µm Flash memory cell (W=0.25µm, L=0.375µm), derived from (1) calculating V_{FG} using the charge balance model.

The charge balance model has been used also to investigate the bias-dependence of capacitive coupling coefficients. Such coefficients have been derived from (1), after the FG voltage has been calculated by means of the charge balance model assuming $Q_{FG}=0$. Figures 3-8 and 3-9 show the CG and D coupling coefficients of a 0.25µm Flash memory cell evaluated in bias conditions corresponding to usual program and read operations: $V_S=0V$ and $V_B=-1V$, while V_{CG} and V_D range from 0V to 9V and from 0V to 5V, respectively. As shown in Figure 3-8, α_{CG} does not depend significantly on D bias, whereas it reduces abruptly when V_{CG} is increased above 2-3V, i.e. moving from subthreshold to above-threshold conditions [15].

Even if the variation of this capacitive coefficient is rather small (6-7 percent), the error in the V_{FG} evaluation occurring when $V_{CG} >8$ can be larger than 0.4V. Further, since program/erase FN currents depend exponentially on the oxide field, i.e. on the FG potential, the wrong V_{FG} estimate can determine errors as large as some order of magnitude in the FN current calculation. Definitely, this demonstrates the voltage dependence of the CG coefficient, whose physical reasons relying on the V_{CG} dependence of the FG to B capacitance have been exhaustively explained in Paragraph 1.3.2. A further example demonstrating the bias dependency of capacitive coupling ratios is given by Figure 3-9, showing the α_D trend versus CG and D voltages. Similarly to α_{CG}, the D coupling coefficient reduces passing from subthreshold to above-threshold conditions, whereas it rises strongly if V_D is reduced below 1-2V when $V_{CG}>3.5V$. Such a bizarre behavior versus V_{CG} and V_D is related to strong increase of the FG-D capacitance (see Paragraph 1.3.2).

Finally, the charge balance model has been tested in more complex circuit simulations. To this regard, a circuit developed to perform the read operation of EEPROM memory cells has been considered. As the circuit comes from a Smart Card application, the sensing scheme is accurate in terms of current detection, it ensures large sensitivity in differentiating programmed and erased states, and at the same time it is fast enough to work with the micro controller clock speed [26]. The sensing scheme is also reliable in terms of disturbs and cycling.

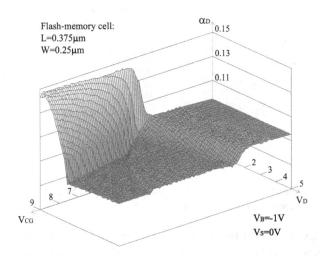

Figure 3-9. D coupling coefficient of a 0.25μm Flash memory cell (W=0.25μm, L=0.375μm), derived from (1) calculating V_{FG} using the charge balance model.

Figure 3-10 shows the block representation of sensing circuits, namely the read path. The sensing scheme generates the voltages to bias the Bit Lines (BL) and Control Gate (CG) of the cells in the array. BL voltage should be large to drive enough current in the cell, but also small to avoid any stress effect during read. The CG voltage (V_{CG}) is generated to bias a virgin cell to have a current level between erased and programmed states. V_{CG} is dynamically applied to the CG only during read, to eliminate any unwanted disturb. A virgin cell is therefore used to generate the CG voltage, and therefore a CM is needed to perform correct simulations.

The schematic of the sense amplifier circuit simulated is shown in Fig. 3-11. It is a classic scheme where active load p-channel transistors are biased to provide the wanted constant current, thus allowing a controlled trip point voltage and temperature compensation [26]. The structure is fully differential to have good noise immunity. Mn1 and Mn3, Mn2 and Mn4 provide the current/voltage conversion to bias the reference cell and the cell to be read in the matrix. V_{CELL} and V_{REF} are voltages deriving from the I-V conversion of currents driven by the cell in the memory array and the reference cell, that are compared to generate the V_{SENSE_OUT} digital level.

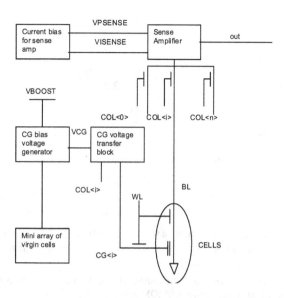

Figure 3-10. Block diagram of the read path circuits implemented in [26] and used in simulations.

Simulation results depicted in Figures 3-12(a)-(b) show that the output signal of the sense amplifier switches correctly according to the programmed/erased state of the EEPROM memory cell. Thus, the charge balance model demonstrates to be effective to simulate FG memory cells also in complex circuits, and therefore it can be used to simulate any circuit including a FG memory cell: read paths, non-volatile latches, X and Y decoders, voltage pumps.

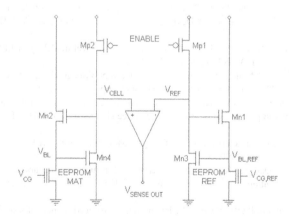

Figure 3-11. Schematic of the sense amplifier and the direct I-V conversion circuits of the EEPROM memory considered, that has been used for circuit simulation.

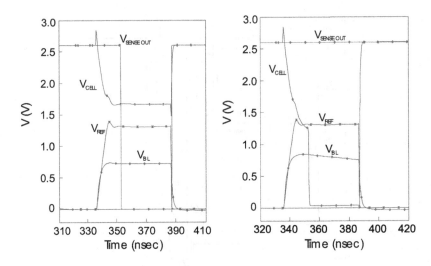

Figure 3-12. Control signals, and sense amplifier output obtained from read-path circuit simulations in the two cases of a programmed (a) and erased (b) E^2PROM memory cell.

REFERENCES

1. 1. P. Pavan, R. Bez, P. Olivo, E. Zanoni, "Flash memory cells – An overview", Proc. of the IEEE, vol. 85, N. 8, pp.1248-1271, 1997.
2. A. Kolodny, S. T. K. Nieh, B. Eitan, and J. Shappir, "Analysis and modeling of Floating Gate EEPROM cells," IEEE Trans. Electron Dev., Vol. ED-33(6), pp.835-844, 1986.
3. J. Suñè, M. Lanzoni, R. Bez, P. Olivo, and B. Riccò, "Transient simulation of the erase cycle of floating gate EEPROMs," in IEDM Tech. Dig., pp.905-908, 1991.
4. S. Keeney, R. Bez, D. Cantarelli, F. Piccinini, A. Mathewson, L. Ravazzi, C. Lombardi, "Complete Transient Simulation of Flash EEPROM Devices", IEEE Trans. Electron Dev., Vol. ED-39(12), pp. 2750-2757, 1992.
5. M. Lanzoni, J. Suné, P. Olivo, B. Riccò, "Advanced electrical-level modeling of EEPROM cells", *IEEE Trans. Electron Dev.*, Vol. ED-40(5), pp.951-957, 1993.
6. L. C. Liong, and P. Liu, "A theoretical model for the Current-Voltage characteristics of a Floating-Gate EEPROM cell," IEEE Trans. Electron Dev., Vol. ED-40(1), pp. 146-151, 1993.
7. S. Chung, C.-M. Yih, S. S. Wu, H. H. Chen, and G. Hong, "A Spice-compatible Flash EEPROM model feasible for transient and program/erase cycling endurance simulation," in IEDM Tech. Dig., pp.179-182, 1999.
8. S. T. Wang, "On the I-V characteristics of floating-gate MOS transistors," IEEE Trans. Electron Devices, vol. 26, no. 9, pp. 1292-1294, 1979.
9. Woong L. Choi and Dae M. Kim, "A new technique for measuring coupling coefficients and 3-D capacitance characterization of Floating-Gate devices," IEEE Trans. Electron Dev., Vol. ED-41(12), pp.2337-2342, 1994.
10. R. Duane, A. Concannon, P.O'Sullivan, A. Mathewson, "Advanced numerical modelling of non-volatile memory cells", Proc. ESSDERC, pp. 304-307, 1998.
11. R. Bez, E. Camerlenghi, D. Cantarelli, L. Ravazzi, and G. Crisenza, "A novel method for the experimental determination of the coupling ratios in submicron EPROM and Flash EEPROM cells," IEDM Tech. Dig., 1990, pp. 99-102.

12. K. Tamer San, Çetin Kaya, David K.T. Liu, Tso-Ping Ma, and Pradeep Shah, "A new technique for determining the capacitive coupling coefficients in Flash EPROM's," IEEE Electron Device Lett., Vol. EDL-13(6), pp.328-331, 1992.

13. M. Wong, D. K.-Y. Liu, and S. S.-W. Huang, "Analysis of the subthreshold slope and the linear transconductance techniques for the extraction of the capacitance coupling coefficients of Floating-Gate devices," IEEE Electron Device Lett., Vol. EDL-13(11), pp.566-568, 1992.

14. B. Moison, C. Papadas, G. Ghibaudo, P.Mortini, and G. Pananakakis, "New method for the extraction of the coupling ratios in FLOTOX EEPROM cells," IEEE Trans. Electron Dev., Vol. ED-40(10), pp.1870-1872, 1993.

15. L. Larcher, P. Pavan, L. Albani, and T. Ghilardi, "Bias and W/L dependence of capacitive coupling coefficients in Floating Gate memory cells," IEEE Trans. Electron Dev., Vol. ED-48(9), pp. 2081-2089, 2001.

16. L. Larcher, P. Pavan, S. Pietri, L. Albani, A. Marmiroli, "A New Compact DC Model of Floating Gate Memory Cells Without Capacitive Coupling Coefficients," IEEE Trans. on Electron Devices, Vol.49, N.2, pp. 301-307, 2002.

17. "The MOS model, level 903", available online:
http://www.semiconductors.philips.com/Philips_Models/

18. " The MOS model, level 1101," available online:
http://www.semiconductors.philips.com/Philips_Models/

19. BSIM3v3.2. Manual, 1999 U. C. Berkeley, USA, 1999. [available online: http://www-device.eecs.berkeley.edu/~ptm]

20. W. Liu, X. Jin, K. M. Cao, an C. Hu, "BSIM4.0.0 MOSFET Model User's Manual," 2000.

21. C. Enz, F. Krummenacher, and E. Vittoz, "An analytical MOS transistor model valid in all regions of operations and dedicated to low voltage and low current applications," Analog Integrated Circuit & Signal Processing, Vol. 8, pp. 83-114, July 1995.

22. M. Bucher, C. Lallament, C. Enz, F. Theodoloz, and F. Krummenacher, "The EPFL-EKV MOSFET model equations for circuit simulation, Version 2.6," Technical Report, Electronics Laboratory, Swiss Federal Institute o Technology (EPFL), June 1997. [available online: http://legwww.epfl.ch/ekv/]

23. G. Gildenblat, N. Arora, R. Sung, and P. Bendix, "Scalable surface potential based compact MOSFET model," Proc. 1997 International Semiconductor Device Research Symposium, p. 33, 1997.

24. T. L. Chen and G. Gildenblat, "Analytical approximation for the MOSFET surface potential," Solid-State Electronics, Vol. 45, pp. 335-339, 2001.

25. http://home.hiroshima-u.ac.jp/usdl/HiSIM.html

26. A. Conte, "High speed and high resolution sensing scheme for embedded EEPROM memories", Proc. of IEEE Non-Volatile Semiconductor Memory Workshop, Monterey (CA) USA, pp. 35-47, February 2000.

Chapter 4

TRANSIENT CONDITIONS: PROGRAM AND ERASE
Theory, compact modeling and circuit simulations

This chapter deals with the task of modeling program and erase operations of FG memory devices. After a brief review of the models proposed in the past in the literature, we will focus on the charge balance model presented in the Chapter 3. Particularly, we will discuss how this model can be extended to simulate program and erase operations of Flash and EEPROM memory cells. This can be done by adding to the basic framework of the DC model a suitable set of voltage-controlled current sources implementing the compact formula modeling program/erase currents of FG memory cells. In this respect, the extension of the charge balance model to account for program/erase simulations reduces to the insertion into the model of a set of current generators.

The development of appropriate CMs reproducing FG memory program/erase currents will be addressed in details considering separately Fowler-Nordheim, Channel Hot Electron (CHE) and CHannel Initiated Secondary ELectron (CHISEL) currents. For each one of these Program/Erase (P/E) currents, the specific difficulties encountered and the approximations taken will be discussed. Simulation results will be compared to experimental measurements to test the validity and the accuracy of the specific P/E current model developed.

1. MODELS PROPOSED IN THE LITERATURE

The modeling of P/E transients of FG memory cells (and particularly of EEPROM cells) has been subject of several papers in the literature [1-10].

One of the first papers on this subject was proposed by Kolodny *et al.* to analyze theoretically and experimentally program-erase characteristics of EEPROM memory cells [1]. To gain insight into basic FG device operations, a simplified CM based on the concept of capacitive coupling coefficients was proposed to simulate program and erase operations. To this regard, a consistent part of the work was devoted to analyze and model physical phenomena occurring during erase, i.e. deep depletion and substrate hole's current, focusing on understanding how such phenomena affect EEPROM memory operations. The analysis and the modeling approach presented in that paper contributed significantly to improve the understanding of FG device operations, and for these reasons this paper can be considered certainly one of the most significant works proposed in the literature on this subject. Nevertheless, the device model proposed therein is based on a FG voltage calculation procedure using the capacitive coupling ratio method, which has been demonstrated to suffer of accuracy limitations that are not negligible (as demonstrated in the previous Chapter).

Successively, a compact SPICE-like model of EPROM memory cell was proposed by Gigon in [2]. Although also this model relied on coupling ratios to evaluate V_{FG}, it featured many interesting characteristics:

1. it was comprised of several common circuit elements (a MOSFET transistor with two gates, i.e. CG and FG; a bipolar transistor; three resistances; two current generators), and therefore its implementation in Spice-like circuit simulators is strongly simplified;
2. it took into account impact ionization (avalanche multiplication current) and parasitic bipolar effects (snap-back);
3. it included Channel Hot Electron current, implementing the lucky-electron model through a current generator.

Further, this model was actually implemented in a commercial Spice-like circuit simulator, demonstrating to be a viable solution for CMs of FG devices. However, although this work constituted certainly a significant improvement in the contemporary research scenario, the model proposed suffered of limits deriving from coupling ratio method used to calculate V_{FG}.

Transient simulations of EEPROM memory erase operation have been the subject of other papers proposed in the literature [4,6]. In [4], a CM comprised of some circuit elements (an MOS transistor, a FN tunnel current injector, and some coupling capacitors) was proposed to simulate program and erase operations of EEPROM memory cells. In particular, the paper focused on two relevant aspects related to the erase operation modeling:

1. deep depletion (non steady-state conditions) occurring in erase when the drain region is not highly doped;
2. electron energy quantization effects at the Si/SiO_2 interface.

Both these phenomena that have to be taken into account for a correct evaluation of the tunnel current, i.e. for the correct simulation of EEPROM write operations were modeled using the FN tunnel injector. Adopting the same basic CM framework, a more detailed analysis of charge quantization effects was carried out in [6]. Particularly, two main effects of charge quantization phenomena on the FN tunnel current were identified and compactly modeled (see Paragraph 3):

1. the surface potential is enhanced by the electron quantization phenomena, and therefore the oxide field reduces considerably compared to that classically calculated;
2. tunneling electrons see an oxide potential barrier which is reduced compared to its classical value: for this reason, to calculate coefficients of the FN current, an effective field-dependent oxide barrier height has to be used in place of the classical constant barrier.

Nevertheless, despite of the great accuracy deserved to the tunnel current modeling, CMs of EEPROM memory cells proposed in [4,6] calculate the FG voltage using the capacitive coupling ratio method, whose accuracy limits degrade simulation quality, making the efforts spent to model charge quantization phenomena ineffective.

In the past, other simulation methodologies have been followed to simulate program/erase operations of FG memory devices.

A 2-D device simulator allowing simulating both program and erasing operations of Flash EEPROM memories was developed in [3,5]. FN and CHE currents through the oxide and Band-To-Band (BTB) tunneling current in the silicon were incorporated in the device simulator through physical models. Although the simulation accuracy was improved compared to previous FG device models, this modeling approach was not suited for circuit simulation purposes. Nevertheless, models developed in device simulator are useful to gain insight into specific physical mechanisms of FG device operations, which cannot be investigated in depth using CMs implemented in circuit simulators [11].

The different role of device simulator models was confirmed again in [8], where the modeling of the erase operation has been addressed through a 3-D device simulator incorporating FN and BTB tunnel currents. Since this model is particularly expensive from the computational point of view, it is not suited for circuit simulations, and it can be used to study in details physical mechanisms of FG device operations.

Recently, other CMs of both EEPROM and Flash memory cells have been proposed in the literature [7,9,10]. In [7], a macro-model of EEPROM memory cells has been developed in a Spice-like circuit simulator, including drain and tunneling current and employing the capacitive coupling ratio method to evaluate the FG voltage. In [9], a complete DC and transient model of Flash EEPROM memories has been presented. This model incorporated both program CHE current (the lucky-electron model is implemented) and the erase channel FN current. Further, it allowed simulating P/E cycling endurance characteristics, i.e. the threshold voltage window closure by modeling in a simple way interface state generation due to CHE current and charge pile up within the oxide. However, even though the surface potential drops at CG, D, S and B oxide interfaces are taken into account, also the V_{FG} calculation procedure of this model is based on capacitive coupling ratios, with the known consequences on the simulation result accuracy.

The same drawback is featured also by the EEPROM macro model recently proposed in [10], which describes accurately FN program/erase currents modeling in details the surface potential drop at the drain junction.

In conclusion, papers summarized above give a wide range of possible solutions for the compact modeling of FG memory devices. To this regard, few points concerning all the above mentioned CMs are worth to be underlined:

1. they rely on constant capacitive coupling coefficients method to evaluate the FG voltage, thus suffering of non negligible accuracy limits;
2. the above CMs represent a different tradeoff between simulation accuracy and implementation complexity, but no original solutions have been proposed till now to overcome the limits of the capacitive coupling ratio method [1].

For this reason, a new modeling approach is necessary to overcome such limits, guaranteeing improved performances without excessive complexity and making very fast and easy the use, application and upgrade of FG device models.

2. THE CHARGE BALANCE MODEL: THE EXTENSION TO TRANSIENT CONDITIONS

To simulate the program/erase operations of FG devices, we have assumed the charge balance model in DC conditions presented in the Chapter 3. To extend this model to account for transient simulations, a suitable set of voltage controlled current sources have been added to the basic framework of the model to implement compact formulae of program/erase currents. The number and the position of current generators depend on the FG memory considered (Flash or EEPROM cells) and the writing mechanisms used to transfer charge to and from the FG.

For example, to extend the DC model of an EEPROM cell to account for transient simulations, a voltage controlled current source has to be inserted between FG and D to model P/E currents. Thus, as shown in Fig. 4-1, the compact DC and transient model of an EEPROM cells will be comprised of four circuit elements: i) the capacitor, C_{CG}; ii) the MOS transistor equivalent to the dummy cell; iii) the voltage-controlled voltage source, V_{FG}, which is necessary in DC conditions to initialize the FG node to its correct voltage; iv) the voltage-controlled current source, I_{TUN}, which models the FN tunnel current flowing across the tunnel oxide between D and FG. This last circuit elements, which constitutes the only difference compared to the DC model, implements the analytical expression of FN currents flowing across the tunnel oxide during program and erase operations.

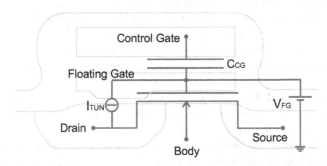

Figure 4-1. The compact charge balance model of an EEPROM memory cell is comprised of: the C_{CG} capacitor; the MOSFET transistor equivalent to the dummy cell; the voltage controlled voltage source, V_{FG}; the voltage controlled current source, I_{TUN}, which models the Fowler-Nordheim current.

Differently, to extend the DC model of Flash memory devices to account for transient conditions, three voltage controlled current sources have to be added to reproduce program/erase currents:
1. a voltage controlled current source has to be inserted between FG and S, I_{W1}, which models the FN current flowing at the source side. This current generator is needed when modeling Flash memories that were erased by FN tunnel at the source side;
2. a voltage controlled current source connected between FG and B, I_{W2}, which models the FN tunnel current flowing toward the substrate;
3. a voltage controlled current source connected between FG and D, I_{W3}, which models CHE and CHISEL injection current, via suitable compact formulae.

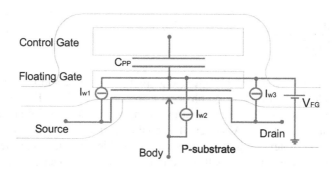

Figure 4-2. The compact charge balance model of a Flash memory cell is comprised of: the C_{CG} capacitor; the MOSFET transistor equivalent to the dummy cell; the voltage controlled voltage source, V_{FG}; three voltage controlled current source, I_{W1}, I_{W2} and I_{W3}, that model P/E currents: FN (I_{W1} and I_{W2}), CHE and CHISEL currents (I_{W3}).

The versatility of the charge balance model to adapt itself to different FG memory cells constitutes another great advantage, among those pointed out in Chapter 3, which promotes its application and use.

When using the charge balance model, the simulation accuracy of FG device program/erase operations depends strictly on the precision of CMs developed to describe FN, CHE and CHISEL currents. Therefore, great attention has to be devoted to develop effective CMs of these currents mechanisms, which will be presented separately in the following Paragraphs along with P/E simulation results obtained.

3. FOWLER-NORDHEIM CURRENT

The well-known formula of Fowler-Nordheim current will be presented along with its basic theory in the first part of this Paragraph, showing approximations and numerical adaptations that have to be assumed for an effective CM of FN current.

In the second Part, simulation results will be compared to experimental data measured on EEPROM and Flash memory cells to prove the fitting capabilities of the model. Moreover, using the charge balance model including the FN current generator, P/E transients of EEPROM memory cell will be carefully analyzed, highlighting limits of the usual experimental method adopted to derive them [12]. Effects of the gate/drain voltage pulse on threshold voltage and tunnel current will be also briefly discussed.

3.1 Theory and compact modeling

The concept of tunneling through a potential barrier stems from the nonzero solutions of the Schroedinger equation in classically forbidden regions [13]. The probability of electron-tunneling depends on the distribution of occupied states in the injecting material and on the shape, height and width of the potential barrier. Using a free-electron gas to model the electron population in the injecting material and the Wentzel-Kramers-Brillouin (WKB) approximation to calculate the tunneling probability [14], the well known expression for FN current density can be obtained [15].

$$J_{FN} = \frac{\pi k T C_{FN}}{\sin(\pi k T C_{FN})} A_{FN} F_{OX}^2 \cdot \exp\left(-\frac{B_{FN}}{F_{OX}}\right) \tag{1}$$

where F_{OX} is oxide field, T is the temperature and A_{FN}, B_{FN} and C_{FN} are coefficients that can be calculated from MOS physical constants [15]. The FN current depends exponentially on F_{OX}, and therefore a strong control on the oxide field is required to assure the desired J_{FN} level. Although this could be trivial, it hides some traps. For example, if a given potential drop is applied to the oxide layer to force the desired F_{OX}, i.e. the desired J_{FN}, a relatively small oxide thickness variation can change strongly F_{OX} and J_{FN}, moving them away from their expected values. Therefore, a very strong control of device manufacturing process against oxide thickness variations is needed to avoid strong J_{FN} variations.

Further, although the FN current formula is well known, its practical use for compact modeling purposes deserves to be briefly discussed.

1. Since the FN current formula is not capable to fit accurately the experimental variations of the FN current against temperature changes, the first factor of Equation (1) is usually neglected [16,17]. Thus, the FN current formula reduces to $J_{FN} = A_{FN} F_{OX}^2 \cdot \exp(-B_{FN}/F_{OX})$, and the temperature dependence of FN current is sometimes accounted for through semi-empirical formula modeling the temperature dependence of the potential barrier height.

2. Even if FN coefficients can be evaluated exactly from physical constants [15], A_{FN} is often assumed as a fitting parameter to adjust simulations and reduce their gap with measurements. It is interesting to note that this practice, which is quite common and is not correct from the theoretical point of view, represents an empirical way to account for differences of device manufacturing processes, that are very difficult to taken into account using the classical FN current formula. For example, process receipt variations among several samples can affect the quality of the oxide and/or the Si/SiO_2 interface, determining oxide potential barriers that are theoretically equal but actually (slightly) different. Since taking into account such differences through standard MOS physical constants is very difficult (practically impossible for circuit simulation purposes), it is preferable to accomplish this task by using A_{FN} as a fitting parameter, or, alternatively, by introducing a multiplication factor which is adjusted to give the best agreement between measurements and simulations.

Classical expressions to calculate FN coefficients from MOS physical constants are given in the following [15].

$$A_{FN} = \frac{q^3 m_{Si}}{16\pi^2 \hbar m_{OX} \Phi_0} \qquad B_{FN} = \frac{4\sqrt{2 m_{OX} \Phi_0^3}}{3q\hbar} \qquad C_{FN} = \frac{2\sqrt{2 m_{OX} \Phi_0}}{\hbar q F_{OX}} \tag{2}$$

q is the electron charge, and $\hbar$ is the reduced Plank's constant; Φ_0 is the oxide barrier height; m_{Si} and m_{OX} are the electron effective mass in the silicon and in the oxide layer, respectively.

Several values have been proposed in the literature for the electron effective mass in the oxide, and there is still no agreement about the correct value that has to be used. As evidenced in [18], the specific value assumed for m_{OX} depends sometimes on the model assumed to solve the Schroedinger equation and to calculate the transmission probability through the oxide barrier. Nevertheless, in agreement with most common values found in the literature [15,18-25], the range of reasonable m_{OX} values should vary between $0.4 \cdot m_0$ and $0.5 \cdot m_0$, where minor m_{OX} differences should be related to different oxidation processes and m_0 is the free electron mass.

In silicon with <100> orientation the electron effective mass can assume two values, that are related to the twofold longitudinal ($m_L = 0.916 \cdot m_0$) and fourfold transverse ($m_T = 0.19 \cdot m_0$) valleys (with respect to the direction orthogonal to the Si/SiO$_2$ interface) where electrons are distributed. Since the fourfold transverse valley is the most populated one in common bias conditions, $m_{Si} = 0.19 \cdot m_0$ is a good approximations for compact modeling purposes. Still, if the cathode is made of polycrystalline silicon, which can be considered to a first order approximation as degenerately doped silicon, m_{Si} has to be slightly increased to $0.3 \cdot m_0$ in agreement to [22].

The oxide barrier height is a well known physical quantity, and for the classical Si/SiO$_2$ interface it is given by $\Phi_0 = 3.12$ eV. Since in modern memory devices control and floating gate are made of polycrystalline silicon, the poly-Si/SiO$_2$ barrier height has also to be estimated. In fact, although it is common practice to use the classical Si/SiO$_2$ barrier height in place of the poly-Si/SiO$_2$ one, the poly-Si/SiO$_2$ barrier height is lower than 3.12 eV, and errors occurring by taking this approximation are not negligible, since $\Phi_0^{1.5}$ is a multiplication factor of the exponential coefficient of FN current formula. The reduction of the poly-Si/SiO$_2$ barrier height, which can be reasonably assumed as $\Phi_0 \approx 2.9$ eV [26], are mainly due to the enhancement of the local electric field generated by the increased roughness of the poly-Si/SiO$_2$ interface (compared to Si/SiO$_2$ one), and the local charge piled up at the boundaries between different polycrystalline silicon grains.

The physical picture depicted above is certainly true when considering the classical theory, but it does not hold any more when charge quantization phenomena are taken into account. Particularly, charge quantization phenomena, that cannot be neglected any more to model correctly the physics of modern VLSI device, affect significantly both the oxide barrier height and the electric field within the oxide, complicating considerably their calculation [4].

3.1.1   Charge quantization effects on oxide barrier height

When the silicon surface is inverted or accumulated (which are the usual conditions during tunnel injection in MOSFET and FG memories), carriers are confined into a narrow potential well, so that their energy in direction perpendicular to the interface is quantized.

Therefore, contrarily to the classical theory which considers electrons as a three-dimensional gas of free-particles with a Boltzmann energy distribution, the correct physical treatment is to view them as a two dimensional quantum-mechanical gas [27]. Since the carrier energy in the direction normal to the oxide interface is quantized, carriers are distributed in discrete energetic sub-bands within the conduction band. Further, according to a proper quantum mechanical description carriers move some distance away from the silicon/oxide interface, so that the charge centroid is no more located a the silicon/oxide interface, as predicted by the classical theory [23,28-33].

Major effects of charge quantization phenomena on FN current modeling can be grouped schematically in two sets, related to F_{OX} and Φ_0 calculations, respectively.

1. Since the charge centroid is located at some distance from the oxide interface, the electrical thickness of the oxide layer becomes greater than the physical oxide thickness, thus leading to a gate capacitance reduction and to a surface potential enhancement with respect to those predicted by the classic theory. As it will be shown in the next Paragraph, this determines a considerable reduction of the oxide field compared to that classically estimated, so that a proper modeling of quantum phenomena is necessary to correctly compute F_{OX} for FN current calculation purposes.

2. Since the carrier energy is quantized in the direction normal to the interface, electrons are not concentrated at the bottom limit of the conduction band, E_C, but they are distributed in discrete sub-bands whose energy levels are higher than E_C. Therefore, tunneling electrons see a reduced oxide potential barrier compared to the classical Φ_0. For this reason, as the oxide barrier height influences the FN coefficients calculation, an *effective* barrier height, Φ_E, has to be introduced in place of Φ_0 (see Fig.4-3). Φ_E is defined as the difference between Φ_0 and the energy level of the lowest sub-band measured from the bottom of the silicon conduction band, E_0 [4,6].

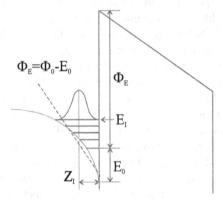

Figure 4-3. Schematic picture of charge quantization effects at the silicon-oxide interface. E_I and Z_I are the energy level and the centroid distance from the interface of the Ith quantized charge sub-band. Φ_E is the *effective* height of the oxide potential barrier.

However, E_0 is not simple to evaluate, as the theoretical treatment of quantum phenomena requires a complex mathematical apparatus to solve self-consistently Schroedinger's and Poisson's equations. Such equations are usually solved through numerical methods that are not suited for implementation into CMs. Anyway, provided that some inaccuracy in the E_0 estimate can be tolerated, an analytical expression for the ground quantized energy level, E_0, can be achieved approximating the exact electrostatic potential at the Si/SiO$_2$ interface by a linear potential (dotted lines) [23,31-33].

$$E_0 = \frac{A_0 \left(\hbar q F_{Si}\right)^{2/3}}{\left(2m_{Si}\right)^{1/3}} \qquad (3)$$

F_{Si} is the effective electric field for carriers in the silicon [31,34], and $A_0 = 2.338$ is the first zero of the Airy's function.

As E_0 depends on F_{Si}, the *effective* barrier height is voltage dependent [35]. Thus, by replacing Φ_E with Φ_0 to calculate FN coefficients in (2), we obtain a FN current formula, where A_{FN} and B_{FN} are functions of the electric field [6].

3.1.2 The oxide field calculation

Although the exact knowledge of F_{OX} is necessary for a correct simulation of the FN current, the oxide field is generally calculated by approximate methods often neglecting charge quantization and poly depletion effects. As anticipated in the previous Paragraph, charge quantization and poly depletion enhance surface voltage drops at both silicon and polycrystalline silicon/oxide interfaces, thus determining the reduction of the oxide field compared to the classically calculated one.

Unfortunately, to model accurately charge quantization phenomena, numerical methods solving self-consistently Schroedinger's and Poisson's equations have to be employed, and CMs can be adopted only if some inaccuracies can be tolerated.

Generally, F_{OX} can be calculated solving the following equation, that equates the voltage applied between electrodes at the two sides of the oxides, V, to the sum of voltage drops across the oxide, $F_{OX} \cdot T_{OX}$, and at cathode and anode oxide interfaces, ψ_C and ψ_A, that are zero for metal cathode and anode [4,34,36].

$$V = F_{OX}T_{OX} + \psi_C + \psi_A + V_{FB} \qquad (4)$$

T_{OX} is the oxide thickness. V_{FB} is the flat band voltage, and it is given by $V_{FB}=\Phi_{MS}-Q_{PA}/C_{OX}$. Φ_{MS} is the work function difference between cathode and anode materials; Q_{PA} models all the parasitic charges per unit area at the Si/SiO$_2$ interface; $C_{OX}=\varepsilon_{OX}/T_{OX}$ is the oxide capacitance per unit area, where ε_{OX} is the insulator permittivity.

The key point to evaluate F_{OX} is to calculate ψ_C and ψ_A as a function of F_{OX}, so that (4) becomes a simple equation in one unknown term, i.e. F_{OX}.

Depending on the physical accuracy degree, which influences directly the complexity of ψ_C and ψ_A expressions, the solution of this equation can be obtained either in analytical way (a closed form formula) or using numerical/iterative methods.

In the following, three different options differing for the tradeoff between accuracy and complexity will be briefly presented to evaluate F_{OX}, going from the most approximated solution to a calculation method including both charge quantization and poly depletion effects.

3.1.2.1 The simplest F_{OX} calculation method

To perform a fast and approximated calculation of the electric field within the oxide, we have taken some simplifying assumptions regarding cathode/anode operation conditions.

If the cathode/anode is in accumulation, the surface potential drop at the oxide interface can be neglected, so that $\psi_{C/A} = 0$. This hypothesis is as much more accurate as the cathode/anode material doping is higher, and generally it leads to very small errors in the F_{OX} evaluation.

If the cathode/anode is in depletion conditions, the surface potential drop can be calculated through Eq. (5), which has been derived from the Gauss's law applied at the Si/SiO$_2$ interface.

$$\psi_{C/A} = \frac{\varepsilon_{OX}^2}{2qN_{C/A}\varepsilon_{Si}} F_{OX}^2 \tag{5}$$

ε_{Si} is the silicon permittivity; q is the electron charge; $N_{C/A}$ is the doping of the cathode/anode material. Since $\psi_{C/A}$ is inversely proportional to $N_{C/A}$, the surface potential drop at the oxide interface can be neglected when the cathode/anode is made of either polycrystalline silicon (CG and FG) or highly doped silicon (D and S), provided that the bias applied were low. Thus, $\psi_{C/A} = 0$ can be assumed provided that some errors leading to a considerable F_{OX} overestimate can be tolerated. To this regard, note that such errors increase proportionally to V, so that $\psi_{C/A} = 0$ leads to an inaccurate F_{OX} estimate in erase/program conditions of FG memories (large V values).

If the cathode/anode is inverted, the surface potential drop can be fixed to $\psi_{C/A} = 2\psi_F + 6\psi_T$. ψ_T is the thermal voltage, and ψ_F is the cathode/anode Fermi potential, which is given by $\psi_F = \psi_T \cdot \ln(N_{C/A}/n_i)$, where n_i is the silicon intrinsic carrier concentration [36].

Applying the above hypotheses, the oxide field calculation is strongly simplified, as F_{OX} results from the solution of an equation, which is at most of the second order. Thus, the computational complexity is strongly reduced, even though there can be some discontinuities between adjacent operation regions (accumulation-depletion-inversion) in the F_{OX} solution, that can generate numerical convergence problems. In conclusion, regardless of the approximations taken, this method is an advantageous tradeoff between complexity (an analytical formula in closed form has been obtained) and results accuracy.

3.1.2.2   The classic theory solution

If the classic theory is applied, the electric field within the oxide is calculated solving the following equation, that comes from the application of the Gauss's law at cathode/anode interfaces with the oxide.

$$F_{OX}\varepsilon_{OX} = Q_{C/A}(\psi_{C/A}) \tag{6}$$

$Q_{C/A}$ is the cathode/anode charge per unit area, which is function of the cathode/anode surface potential drop according to the classical theory [36].

$$Q_{C/A} = \pm\sqrt{2\varepsilon_{Si}qN_{C/A}} \cdot$$
$$\cdot \sqrt{\psi_T \cdot e^{\left(\frac{-\psi_{C/A}}{\psi_T}\right)} + \psi_{C/A} - \psi_T + e^{\left(\frac{-2\psi_F}{\psi_T}\right)}\left[\psi_T \cdot e^{\left(\frac{\psi_{C/A}}{\psi_T}\right)} - \psi_{C/A} - \psi_T\right]} \tag{7}$$

For an n-doped silicon layer, $\psi_{C/A}$ has to be considered negative when the cathode/anode is in accumulation conditions. Equation (7) holds for both n an p type silicon layers regardless the cathode/anode operation conditions (accumulation, depletion, inversion), and it can be used also to model the surface charge density of p+ and n+ polycrystalline silicon layer, thus accounting in a simple way for polycrystalline silicon depletion effects.

Unfortunately, since the $Q_{C/A}$ dependence on $\psi_{C/A}$ is quite complex, it is impossible to achieve an analytical solution of Equation (6), i.e. to express $\psi_{C/A}$ as a function of F_{OX} through closed form formula. For this reason, contrarily to the previous case, numerical methods have to be used to calculate F_{OX}.

In conclusion, compared to the previous calculation method, the main advantages of the classic procedure are:

1. the accuracy of the F_{OX} estimate is improved;
2. the F_{OX} solution does not have discontinuity between different operation regions, thus avoiding possible convergence problems.

The price to pay for the improvement of the F_{OX} estimate accuracy is the increase of the computation complexity, which is quite relevant and makes not convenient the use of this F_{OX} calculation method in comparison to the previous one.

3.1.2.3 The F_{OX} calculation including charge quantization effects: an approximated solution

As explained in previous paragraphs, charge quantization phenomena at Si/SiO$_2$ interface have to be taken into account to calculate accurately the oxide field when device sizes approach VLSI-ULSI scenario. Particularly, a proper modeling of charge quantization effects is necessary to evaluate accurately $\psi_{C/A}(F_{OX})$ not only in inversion, but also in accumulation provided that the doping of the silicon cathode is lower than $\sim 10^{19}$ cm^{-3} [34].

As explained above, if charge quantization occurs, carriers in accumulation and inversion conditions are distributed in quantized energy sub-bands and move some distance away from the silicon/oxide interface, so that the surface potential is enhanced compared to that classically calculated.

Even though the rigorous theory of quantum phenomena is a very complex mathematical problem requiring the self-consistent solution of Schrödinger's and Poisson's equations (that can be achieved only using numerical techniques), some hypotheses can be assumed to simplify the theoretical treatment. In this respect, by approximating the exact electrostatic potential at the oxide interface by a linear potential, the quantized energy levels, $E_{i,j}$, and the charge centroid in the direction normal to the surface, $Z_{i,j}$, can be expressed by simple analytical expression [23,31-33].

$$E_{i,j} = \frac{\left(\hbar q F_{Si}\right)^{2/3}}{\left(2m_{z,i}\right)^{1/3}} A_j \qquad (8)$$

$$Z_{i,j} = \frac{2E_{i,j}}{3qF_{Si}} \qquad (9)$$

A_j is the jth zero of the Airy's function, where the index j indicates the sub-band considered. $m_{z,i}$ is the electron/hole effective mass in the direction normal to the surface.

The surface densities of electrons and holes of each sub-band are given by Equations (10) and (11), where the difference between the Fermi level and the electron sub-band energies are E_F-$E_{i,j}$=q$\psi_{C/A}$-E_G/2-qψ_F-$E_{i,j}$ and between the hole sub-band energies and the Fermi level are $E_{i,j}$-E_F=qψ_F-E_G/2-$E_{i,j}$-q$\psi_{C/A}$, respectively.

$$N_{E,i,j} = \frac{m_{D,i}g_i k_B T}{\pi\hbar^2} \ln\left[1+\exp\left(\frac{E_F - E_{i,j}}{k_B T}\right)\right] \tag{10}$$

$$N_{H,i,j} = \frac{m_{D,i}g_i k_B T}{\pi\hbar^2} \ln\left[1+\exp\left(\frac{E_{i,j} - E_F}{k_B T}\right)\right] \tag{11}$$

E_G is the energy bandgap; k_B is the Boltzmann's constant; $m_{D,i}$ is the density-of-states effective mass, and g_i is the degeneracy factor [23,31-33]. For a crystal silicon with <100> orientation, electron effective masses are the usual longitudinal ($m_{Z,1}$=0.916·m_0) and transverse ($m_{Z,2}$=0.19·m_0) masses, whereas the density-of-states effective masses are given by $m_{D,1}$=0.19·m_0 and $m_{D,1}$=0.417·m_0 (the degeneracy factors are g_1=2 and g_2=4) [23,31]. Hole bands have been considered parabolic despite of their warped shape: effective and density-of-states masses for light and heavy holes are given by $m_{Z,1}$=$m_{D,1}$=0.26·m_0 and $m_{Z,2}$=$m_{D,2}$=0.7·m_0, [28,37].

$$N_{E(H)} = \sum N_{E(H),i,j} \qquad Z_{E(H)} = \sum N_{E(H),i,j} Z_{i,j}/N_{E(H)} \tag{12}$$

$N_{E(H)}$ is the total electron (hole) charge density, and $Z_{E(H)}$ is the distance from the cathode/anode oxide interface of the charge centroid.

Since the charge centroid has a non-zero value, the surface potential drop is enhanced compared to that calculated following the classical theory. By assuming the electron and hole charges as sheets of infinitesimal width placed at $Z_{E/H}$ distance from the oxide interface, the increase of the surface potential drop due to charge quantization can be easily calculated.

$$\Delta\psi_Q = \frac{qN_{E/H}(\psi_{C/A})Z_{E/H}(\psi_{C/A})}{\varepsilon_{Si}} \tag{13}$$

However, $N_{E/H}$ and $Z_{E/H}$ themselves are function of the surface potential drop, which is given by $\psi_{C/A}$=ψ_{CL}+$\Delta\psi_Q$, where ψ_{CL} is the surface potential evaluated following the classical theory. In other words, this means that (13) is an equation whose unknown variable is given by $\Delta\psi_Q$. To solve this equation, i.e. to calculate $\psi_{C/A}$, numerical/iterative methods have to be employed, as $N_{E/H}$ and $Z_{E/H}$ feature a complex dependence on the cathode/anode surface potential drop.

Further, since the equation complexity depends strictly on the number of sub-bands considered, the number of sub-bands taken into account has to be limited to reduce the computational complexity. In this respect, a good approximation is to consider only the three lowest energy sub-bands. In fact, errors occurring with this approximation are negligible for CM applications; even though they increase slightly when low silicon fields are considered, being the occupation probability of the highest sub-bands larger for lower F_{OX} [23,34].

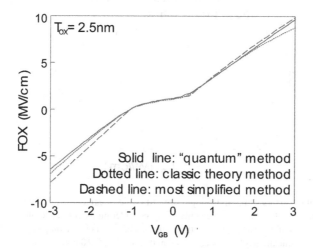

Figure 4-4. Oxide field calculated on a n+polysilicon/p-silicon MOSFET capacitor (T_{OX}=2.5 nm) using the three methods discussed in the text: 1) the most approximated one (dashed line); 2) the one derived directly from the classic theory (dotted line); 3) that including charge quantization effects (solid line).

Compared to previous F_{OX} calculation methods, the modeling of charge quantization phenomena becomes more important as the oxide thickness gets thinner. In Figure 4-4, results of F_{OX} calculation methods are shown for an MOS capacitor whose oxide thickness has been chosen so thin (T_{OX}=2.5 nm) to emphasize the differences between the different calculation procedures. In fact, F_{OX} calculation errors due to the exclusion (or the inaccurate modeling) of charge quantization effects increase on reducing T_{OX}, since $Z_{E/H}$ becomes a larger fraction of the whole electrical oxide thickness.

This is clearly shown in Fig. 4-5, where the percentage errors occurring using the most approximated and the "classical theory" method to calculate F_{OX} are depicted in accumulation conditions. Errors in F_{OX} estimate have been calculated by considering as correct the results obtained from the third F_{OX} calculation method, i.e. the one including charge quantization effects. As shown and previously anticipated, F_{OX} estimate errors grow significantly on both decreasing T_{OX} (since the charge centroid becomes a larger fraction of the oxide thickness) and increasing F_{OX}, since sub-bands with higher energy levels are more populated at lower field.

Although the modeling of quantum phenomena described above is quite simplified compared to a rigorous theoretical treatment, it features the drawback of a relevant increase of complexity compared to the other two F_{OX} calculation methods.

To conclude, it is worth noting that it is difficult to choose a priori the "best F_{OX} calculation method." As it always happens dealing with CMs, this choice depends on simulation accuracy target and on how much you are willing to pay in terms of computation complexity to reach that target.

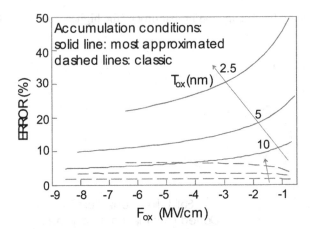

Figure 4-5. Percentage errors of the most approximated (solid lines) and the "classic theory" (dashed lines) F_{OX} calculation methods, that have been calculated by considering as correct results obtained from the F_{OX} calculation method including charge quantization effects. Three n+polysilicon/p-silicon MOSFET capacitors having different oxide thicknesses have been considered: T_{OX}=2.5, 5, 10 nm.

3.2   Simulation results

The introduction of FN current generators into the charge balance model allows simulating both program and erasing operations of EEPROM memories and the erase operations of standard Flash memories.

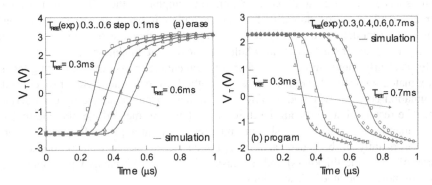

Figure 4-6. Measurements (symbols) and simulations (solid lines) of threshold voltage variations occurring during erase (a) and program (b) operations of a 0.35μm EEPROM memory cell (W=0.3μm, L=0.75μm, C_{CG}=3fF, 7nm tunnel oxide). The threshold voltage is defined as the CG voltage at which I_{DS}=1 μA/μm when V_{DS}=0.8V.

EEPROM memory cells considered in this work have been manufactured in 0.35μm technology: they feature a 7 nm thick tunnel oxide to allow write operations to be performed without applying too high biases, whereas their gate oxide thickness is around 20 nm. To erase this EEPROM memory cell, a voltage ramp from 0 to 12 V is applied to the CG, while S, D, and B are pinned to ground. The program operation is performed applying the same voltage ramp (0 – 12 V) to the D, whereas B and CG are grounded and S is left floating. In both cases a Fowler-Nordheim current flows across the tunnel oxide.

Simulations and measurements of the threshold voltage, V_T, which is defined as the CG voltage at which I_{DS}=1 $\mu A/\mu m$ when V_{DS}=0.8V, are depicted in Figures 4-6(a)-(b) for erase and program operations. As shown, the charge balance model extended to account for FN current simulates excellently threshold voltage variations measured during erase and program operations. Particularly, the fitting quality is excellent independently of rise times of CG and D voltage ramps (the time for the CG ramp to reach the 12 V plateau value) and without any free parameter to adjust simulation results. Using CMs of FG devices it is possible also to monitor physical quantities that cannot be directly measured, like the FG voltage and the FN program/erase currents. In Figures 4-7(a)-(b), tunnel currents simulated in erase and program operations are plotted along with experimental FN currents derived from the measured threshold voltage through I_T=$C_{CG} \cdot dV_T/dt$ [12].

FN current simulations reproduce very well experimental tunnel currents. Particularly, the decrease of I_T maximum on increasing T_{RISE}, i.e. reducing the V_{CG} ramp rate is well simulated by the model. Thus, the model demonstrates to be capable to calculate the FG voltage consistently with the charge injected during P/E operations. In fact, the I_T lowering is due to the fact that with slower V_{CG}-ramps, the greater amount of charge injected (extracted) into (from) the FG compensates more effectively FG voltage variations induced by the CG (D) voltage. Therefore, the FG voltage lowers, and so does the tunnel current.

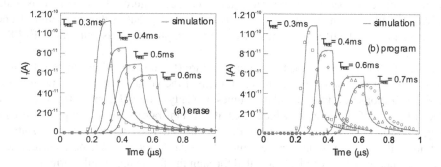

Figure 4-7. Measurements (symbols) and simulations (solid lines) of FN tunnel current flowing through the tunnel oxide (T_{OX}=7 nm) during erase (a) and program (b) operations of a 0.35μm EEPROM memory cell (W=0.3μm, L=0.75μm, C_{CG}=3fF). The experimental FN currents is derived from the measured V_T through I_{TUN}=$C_{CG} \cdot dV_T/dt$

However, although the agreement between I_T measurements and simulations is very good, some discrepancy in the tunnel current simulations can be observed. Precisely, I_T derived from measured data does not reach the same peak value predicted by the model, and it does not flatten in the peak region, as it reduces while the voltage ramp is still rising. Such results, that are apparently in contrast to the same theoretical analysis used to derive them [12], are strictly related to the real shape of voltage ramp applied to the D (CG) to program (erase) the EEPROM memory cell.

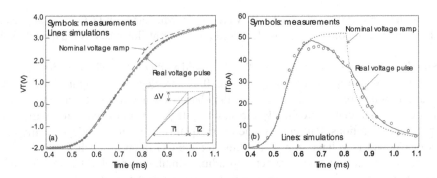

Figure 4-8. Measurements (symbols) and simulations (lines) of the threshold voltage shift (a) and the FN current (b) flowing across the tunnel oxide during the erase of an EEPROM memory cell (W=0.3µm, L=0.75µm, C_{CG}=3fF). Simulations have been performed applying to the CG both the nominal (dotted lines) and the real (continuous lines) voltage ramp (T_{RISE}=0.7ms).

To investigate this point, the real voltage ramp has been measured with an oscilloscope (solid line, in the inset of Fig. 4-8(a)) and, as shown in the same figure, it is quite different from the nominal voltage pulse (dashed line). Then, both V_T and I_T curves during the EEPROM erase considering both the nominal and the real CG voltage ramps have been simulated.

Both threshold voltages and tunnel currents simulated applying the real voltage ramp agree remarkably with measurements, whereas if we use the nominal voltage ramp we cannot simulate properly the experimental data, see Figures 4-8(a)-(b). This indeed confirms the high accuracy of the charge balance model, but at the same time it opens some questions on the usual methods used to analyze program/erase current in EEPROM memory cells.

In fact, following what reported in the literature [12], one would expect that the tunnel current peaks at the end of the ramp, and flattens if the voltage ramp is sufficiently slow. In such conditions, the FG potential assumes a steady-state value since the V_{FG} increase induced by the CG/D voltage (which rises at constant rate) is fully compensated by the charge injected into the floating gate by the FN current. Thus, to maintain such stationary conditions, the constant tunnel current must be equal to the displacement current across the C_{CG} capacitor, which can be calculated through the following simple formula.

$$I_{T,MAX} = C_{CG} \frac{dV_{CG}}{dt} \tag{14}$$

However, as shown in Figure 4-8(a)-(b), the above analysis holds only when considering program/erase currents simulated using the nominal voltage ramp, whereas it fails when it is applied to I_T curves measured and simulated assuming the real voltage pulse. Particularly, the real tunnel current has a lower peak (~10%) compared to that estimated through (14), and its shape versus time does not show any flat region.

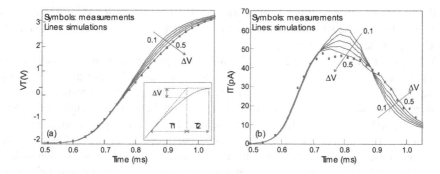

Figure 4-9. Measurements (symbols) and simulations (lines) of the threshold voltage shift (a) and the FN current (b) occurring during the erase of an EEPROM memory cell (W=0.3μm, L=0.75μm, C_{CG}=3fF). Simulations have been performed applying to the CG a real voltage ramp with ΔV ranging from 0.1 to 0.5 V (T_{RISE}=0.7ms).

To investigate this point, the impact of the shape of the CG voltage pulse on V_T and I_T has been studied in details. Particularly, we have considered separately the effects of three parameters which can be used to describe the real voltage pulse: ΔV, T1 and T2, (see the inset of the Figure 4-9(a)). Among them, only the maximum difference between the nominal ramp and the real ramp, ΔV, has a significant influence on both threshold voltage and tunnel currents. Particularly, it influences I_T in two ways (see Figure 4-9(b)): on increasing ΔV, the maximum value of tunnel current reduces significantly (~17% reduction when ΔV=0.5V), and the time interval for which I_T>20 pA increases (~15% increase when ΔV=0.5V).

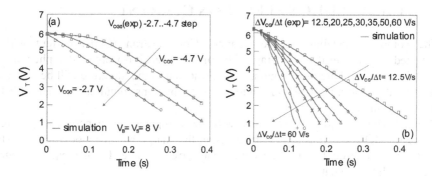

Figure 4-10. Measurements (symbols) and simulations (solid lines) of threshold voltage shifts occurring during the erase of a Flash memory cell (W=0.25μm, L=0.375μm, T_{OX}=10.5nm). The threshold voltage is defined as the CG voltage at which I_{DS}=1 μA/μm when V_{DS}=1.0V. To erase the Flash memory cell a staircase voltage pulse is applied to the CG (V_{CG0} is the initial value)

Using the charge balance model extended to model Flash memory P/E operations (three current generators have been inserted to model tunnel currents flowing across the three oxide portions between FG and S, B, and D, see Figure 4-2), erase transients of 0.25-0.18μm Flash memory cells have been simulated. To erase Flash memory cells with tunnel oxide thickness around 10 nm, a negative staircase voltage pulse is applied to the CG, while S, D, and B are positively biased (~8 V). To test accurately the simulation capabilities of the model, several initial values, V_{CG0}, and rate, $R=\Delta V_{CG}/\Delta T$, of the CG staircase voltage pulse have been considered (ΔV_{CG} is the single CG voltage step, and ΔT is time step of the CG voltage pulse).

As shown in Figure 4-10(a)-(b), threshold voltage shifts simulated and measured during the erase of the Flash memory cell agree excellently regardless V_{CG0} and R differences, and without any free parameter to improve the simulation results.

Since S, D, and B are biased together in erase, tunnel currents through oxide portions above S, D, and B can differ strongly because of substrate doping variations, that impact on the oxide field. Particularly, if surface potential drops at source and drain junction were negligible, tunnel currents flowing toward source and drain wells would be the dominant contribution of the erase current, whereas electrons tunneling to the channel (B) would contribute negligibly to the total erase current due to smaller oxide field. However, this simple assumption does not hold, and since the voltage applied to the gate oxide to erase Flash memory is very large (V_B-$V_{CG}\approx11$-$16V$), surface potential drops at S/D wells can be as large as $2\psi_F$. In this way, F_{OX} differences between oxide portions above D/S wells and B disappear, and although this treatment neglects doping variations carefully modeled in [10], erase current simulations performed fit excellently experimental data, demonstrating that F_{OX} is approximately constant along the oxide interface in such high (V_B-V_{CG}) bias conditions.

In conclusion, note the small oscillations featured by the threshold voltage simulation curves depicted in Figure 4-10(b), that are due to the staircase shape of the erase voltage pulse applied to the CG, and it is not shown by experimental data, that are sampled every ΔT seconds.

4.   CHANNEL HOT ELECTRON CURRENT

The compact modeling of the Channel Hot Electron (CHE) current will be investigated in depth in this Paragraph. The main physical aspects of this phenomenon will be discussed, and the most famous CMs presented in the literature will be illustrated together with approximations and numerical adaptations required for an effective compact modeling.

Channel Hot Electron simulations will be compared to experimental currents measured on MOSFET transistor. Further, using the charge balance model incorporating the CHE current through a voltage controlled current source, Flash memory program operations will be simulated.

In conclusion, difficulties encountered to have an effective CM of the CHISEL current will be briefly discussed, which are mainly due to both the physical complexity and the statistical nature of this phenomenon.

4.1   Theory and compact modeling

The compact modeling of Channel Hot Electron current has been tackled in the literature for a long time.

The most famous CM (and the only one which is completely analytical) of CHE current proposed in the literature was the "lucky-electron" model [38-40]. Despite the many approximations, this model allowed to simulate hot electron injection current through a simple analytical formula, which can be easily included into Spice-like circuit simulators.

Successively, a qualitative model of CHE current was proposed to better understand experimental gate current data measured during the program of Flash memory devices [41]. This model relaxed the weight of impact ionization scattering and considered as spherical the hot electron momentum distribution (contrarily to the "lucky electron" model). Starting from this modeling approach, a physical quantitative CM of CHE injection current will be presented in the second Paragraph of this Section.

4.1.1 The "lucky-electron" model

As explained in Chapter 2, CHE current is composed of those channel electrons passing the potential barrier between the channel and the gate. Qualitatively, to contribute to CHE current, such electrons must gain from the lateral field sufficient kinetic energy to be able to overcome the Si/SiO_2 barrier, and their momentum has to be directed toward the gate for the hot electrons to be collected at the gate.

To gain an intuitive understanding, the physical mechanism of CHE injection has been sketched in Figure 4-11, which represents schematically the hot electron trajectory in the potential-distance space: channel electrons that have gained sufficient kinetic energy from the lateral field in the channel (i.e. *hot electrons*), are redirected toward the gate by a random collision with the crystal lattice. Thus, if such hot electrons do not suffer any collision traveling to the Si/SiO_2 interface, they are swept toward the gate by the oxide field, provided that the gate is at a higher potential than the substrate surface. To quantify the probability that channel electrons contribute to CHE current, i.e. could be collected by the gate, several types of scattering must be taken into account, since they can possibly reduce the amount of electrons arriving at the gate.

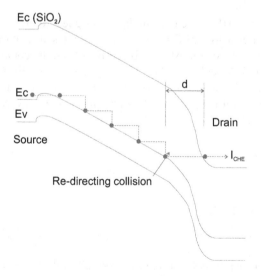

Figure 4-11. Schematic picture of the hot electron path in the potential-distance space.

Physical processes concurring to the CHE injection (electron energy gain, elastic redirection collision, electron traveling from the redirection point to the gate without suffering any inelastic collision) are statistically independent; hence the probability of the whole process is simply given by the product of probabilities of every single process [38].

Assuming a constant accelerating lateral field in the channel, E_C, the "lucky-electron" model is based on the hypothesis that electrons have to travel a distance $d=\Phi_0/q \cdot E_C$ to gain enough kinetic energy to surmount the silicon/oxide barrier. The probability that this occurs is given by $\exp(-d/\lambda)$, where λ (~7-10 nm) is the electron mean free path of hot electrons [38-40]. Following this reasoning and approximating the probability of other scattering events with a constant terms regardless their dependence on the electric field [39], the simple "lucky- electron" formula can be obtained for the CHE current.

$$I_{CHE} = C \cdot I_{DS} \cdot \exp\left(-\frac{\Phi_0}{q \cdot E_M \cdot \lambda}\right) \tag{15}$$

C is a constant, and its value has to be calibrated by measurements. E_M is the peak value of the lateral field at the drain junction. Since I_{CHE} depends exponentially on E_M, the accurate knowledge of the lateral field in the velocity-saturation region is required for a correct calculation of the CHE current.

$$E_M = \frac{V_{DS} - V_{DS,SAT}}{\Delta L} \tag{16}$$

The intuitive interpretation of Eq. (16) is that V_{DS}-$V_{DS,SAT}$ is the voltage drop in the pinch-off region ($V_{DS,SAT}$ is the saturation voltage), while ΔL is the effective length of the velocity-saturation region.

Since the compact CHE current formula is designed to be included in Spice-like models of MOSFET transistor or FG memory devices, the saturation voltage, $V_{DS,SAT}$, and the length of the pinch-off region, ΔL, can be derived from the compact MOSFET model adopted (Philips MOS Model 9/11 [42], BSIM3v3/4 [43,44], EKV [45-46], SP [47,48], HiSIM [49]). Alternatively, simple approximated formulas can be adopted to calculate ΔL and $V_{DS,SAT}$, even though it has been verified that they become as less accurate as device sizes are scaled [50,51].

$$\Delta L = K_L \cdot T_{OX}^{1/3} \cdot X_J^{1/2} \tag{17}$$

$$V_{DS,SAT} = \frac{(V_G - V_T)L_E E_{SAT}}{V_G - V_T + L_E E_{SAT}} \tag{18}$$

X_J is the junction depth; K_L is a constant (=0.22 $cm^{1/6}$ in [52]); L_E is the effective channel length; E_{SAT} is the critical field for velocity saturation (~4·10^4 V/cm in [52]).

4.1.2 Alternative CHE current models

An alternative CM of CHE injection current employing some theoretical concepts of the qualitative model of CHE current proposed in Ref. [41] can be developed to improve simulation accuracy.

This model is based on the following general theoretical concept. To calculate the amount of hot electrons contributing to CHE current, i.e. electrons having kinetic energy higher than the Si/SiO_2 barrier and velocity direction toward the gate, the accurate knowledge of the following physical quantities is needed:

1. the electron energy distribution, which is function of the lateral field;
2. the momentum distribution, to estimate the fraction of electrons directed toward the gate;
3. the probability of an electron with energy E, momentum k and distance d from the Si/SiO_2 barrier to cross the oxide barrier.

1) For the electron Energy Distribution (ED), and particularly to take into account the non-Maxwellian form of its high energy tail, the analytical model proposed in [53] (and used in [54-56]) can be applied. This model is derived on the assumptions that electrons belong to a single non-parabolic conduction band and that their dominant energy loss mechanism is the optical phonon emission.

$$f_{el}(E, E_P) = K \cdot exp\left(-\frac{\chi \cdot E^3}{E_P^\alpha}\right) \qquad (19)$$

E is the electron energy; χ is a constant, whose value has to be determined from Monte Carlo simulations ($\chi = 1.3 \cdot 10^8$ $V^{1.5}cm^{-1.5}eV^{-3}$ in [54]); K is a normalizing constant, and α can vary between 1 [55] and 1.5 [54].

2) Since the time constant of the momentum relaxation is much smaller than the energy relaxation one, i.e. the momentum is randomized well before carriers loose significant amount of their energy, it is reasonable to assume that the momentum distribution is an almost spherical symmetric function ($k_x/k_y \approx 1$) [41].

3) The probability of an electron with energy E and momentum k to cross the oxide barrier between the channel and the gate, P_{IN}, is given by the product of three different contributions:

a) the probability that the electron velocity is directed toward the gate, P_V;
b) the probability that the electron does not loose energy when traveling the distance to the Si/SiO_2 interface, P_C;
c) the probability of electron tunneling through the oxide barrier, P_{TUN}, which becomes one if the electron energy is higher than the oxide barrier height.

a) On the hypothesis that the electron momentum function distribution is spherical, the probability that the electron velocity is directed toward the gate is given by $P_V = \vartheta/4\pi$, where ϑ is the spatial angle in the momentum space directed toward the gate [41], which is assumed constant along the channel.

b) The probability that electrons do not loose energy, i.e. do not suffer inelastic collisions traveling to the oxide interface, can be computed through $P_C = exp(-d/\lambda)$, where d is the distance from the collision re-direction point to the Si/SiO_2 interface. However, for circuit simulation purposes also this probability can be considered constant, and typically $P_C \cdot P_V \approx 0.025$ can be assumed.

c) The electron tunneling probability through the oxide can be calculated applying the WKB method [14], thus obtaining the following classical expression for the FN ($\Phi_0 \leq F_{OX} T_{OX}$) and direct tunneling ($\Phi_0 > F_{OX} T_{OX}$) regime.

$$P_{TUN}[F_{OX}(x)] = \begin{cases} \exp\left(-\dfrac{4\sqrt{2qm_{OX}\Phi_0^3}}{3\hbar F_{OX}(x)}\right) & \text{for} \quad \Phi_0 < F_{OX}T_{OX} \\ \exp\left[-\dfrac{4\sqrt{2qm_{OX}}}{3\hbar F_{OX}(x)}\left[\Phi_0^{3/2} - (\Phi_0 - F_{OX}(x)T_{OX})^{3/2}\right]\right] \end{cases} \tag{20}$$

If the oxide field, F_{OX}, is directed toward the gate, i.e. it is opposed to the electron injection into the gate; the oxide barrier height is increased by the voltage drop across the oxide [41].

To estimate the oxide field along the channel, the knowledge of the channel potential $V_C(x)$ is required, x being the spatial coordinate along the channel (x=0 at the source junction).

$$F_{OX}(x) = \frac{V_G - V_{FB} - V_C(x)}{T_{OX}} \tag{21}$$

$V_C(x)$ can be calculated by integrating the lateral field, $E_P(x)$, which grows almost exponentially from the source to the drain (the increase of the channel field gradient is necessary to support the charge released by the oxide field) [57,58].

$$E_P(x) = E_{MIN} \cosh\left(\frac{x}{L}\right) \tag{22}$$

$$V_C(x) = E_{MIN} L \sinh\left(\frac{x}{L}\right) \tag{23}$$

E_{MIN}, which is the minimum channel field, and L are two parameters, that can be determined solving the following equation system, whose first equation equates the lateral field at the drain junction (x=L_E) to E_M, whereas the second one forces the channel potential drop to be equal to V_{DS}.

$$\begin{cases} E_M = E_{MIN} \cosh\left(\dfrac{L_E}{L}\right) \\ V_{DS} = E_{MIN} L \cdot \sinh\left(\dfrac{L_E}{L}\right) \end{cases} \tag{24}$$

Alternatively, a simpler (but less precise) method to calculate these two parameters is to assume $E_{MIN}=E_{SAT}$ and $L=\Delta L$.

In conclusion, the CHE current can be calculated through the following formula, which computes the fraction of channel electrons able to overcome the oxide barrier by integrating over the energy (to calculate the CHE current at each point along the channel) and over the space (along the channel) the product of the electron energy distribution by the injection probability.

$$I_{CHE} = I_{DS} P_V P_C \int_0^{L_E} \int_0^\infty f_{el}[E, E_P(x)] \cdot P_{TUN}[\Phi_0 - E, F_{OX}(x)] \cdot dE \cdot dx \qquad (25)$$

However, although this formula is quite elegant and easy to understand, some problems could derive from its practical implementation, and particularly from the computational complexity related to the two integrations required to calculate the current. To overcome these limits, some approximations can be applied provided that some inaccuracies can be tolerated.

The first approximation concerns the cancellation of the spatial integral, and it does not determine significant errors. In fact, since the electron energy distribution depends exponentially on the lateral field and E_P grows exponentially approaching the drain junction, the electron energy distribution features a very steep (more than exponential) spatial dependence on x. For this reason, electron distributions computed for $x<L_E$ can be neglected compared to $f_{el}(E, E_P(L_E))$, despite the fact that the F_{OX} inversion point moves from the metallurgical junction inside the channel. Thus, Eq. (25) reduces to a simpler formula, where ΔL_{IN} is a constant representing the equivalent integral distance.

$$I_{CHE} = I_{DS} \Delta L_{IN} P_V P_C \int_0^\infty f_{el}[E, E_P(L_E)] \cdot P_{TUN}[\Phi_0 - E, F_{OX}(L_E)] \cdot dE \qquad (26)$$

Practically, the product $P_V \cdot P_C \cdot \Delta L_{IN}$ can be replaced by a unique multiplication constant, K_{CHE}, that can be used as a fitting parameter to calibrate the simulations. As shown in Fig. 4.12, by choosing properly K_{CHE}, the I_{CHE}-V_G curve plotted considering Eq. (26) is identical to the Eq. (25) one, thus demonstrating that Eq. (26) is capable to reproduce accurately the CHE current without paying the large computational cost of Eq. (25).

The second approximation which can be taken to simplify the CHE current formula concerns the cancellation of the tunneling probability factor. In this way, only the electrons having energy higher than the oxide barrier height (whose probability to cross the barrier is one) are taken into account to calculate the CHE current [41].

$$I_{CHE} = I_{DS} K_{CHE} \int_{\Phi_0}^\infty f_{el}[E, E_P(L_E)] \cdot dE \qquad (27)$$

Although this approximation reduces computational and implementation costs of CHE formula as the tunnel probability calculation is not required, it leads to significant errors increasing at high V_G. Particularly, as shown in Fig. 4.12, and also reasonably expected, neglecting the tunnel probability terms makes the simulated CHE current approaching the Lucky Electron Model (LEM) one, thus making ineffective the use of Eq. (27) compared to the classical LEM formula, which is analytical and very easy to implement.

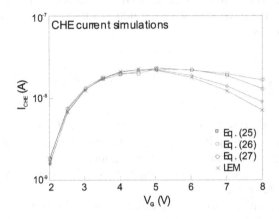

Figure 4-12. CHE current simulations performed adopting compact formulas illustrated in the above paragraph and the Lucky Electron Model (LEM) one (V_{DS}=4.2V).

As shown in Fig. 4.12, as more approximations are introduced in the I_{CHE} calculation, as the simulated CHE current decreases at high V_G assuming the classical "bell" shape, which was a typical feature of CHE current in long channel MOSFET [41]. However, in modern VLSI devices CHE current does not drop off on increasing V_G, but it continues slightly to rise also at high V_G. This is due to the larger oxide field in the drain region of scaled devices, which enhances the tunnel probability thus compensating the strong reduction of the high energy tail of the hot electron distribution induced by the lateral field decrease occurring on increasing the gate voltage. For this reason, the tunnel probability cannot be neglected for an accurate modeling of CHE current.

4.2   Simulation results

By incorporating the CHE model described above into the FG memory device CM (by adding to its basic structure a voltage-controlled current source), quite accurate simulations of threshold voltage (V_T) shift occurring during Flash memory program can be obtained.

To this regard, Figure 4.13 show the threshold voltage shift curves measured (symbols) and simulated (lines) during the program operation of a 0.18 µm Flash memory cell. For Flash memories used here, program operation is performed by applying to the Control Gate the voltage ramp plotted in the inset of Figure 4-12, when source and body are grounded and the drain is biased at 4.2 V.

The CHE formula described by Eq. (26) is implemented into the current generator inserted into the FG CM. Thus, the measured V_T shift occurring during Flash program is well simulated by the model without any additional parameter to improve the fitting

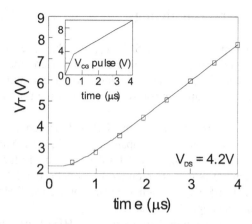

Figure 4-13. Measurements (symbols) and simulations (solid lines) of threshold voltage shift occurring during the program of a 0.18 μm Flash memory (L=0.3 μm, W=0.16 μm).

quality, thus confirming that the this model constitute an effective tool to predict accurately FG device operations.

Even by using the less accurate LEM formula into the current generator included in the FG model, simulation results agree accurately with experimental curves (see Figure 4-14). Figure 4-14 shows the final threshold voltage (V_{TF}) simulated and measured after the Flash memory cell has been programmed. Except minor discrepancies, really not important considering the circuit simulation applications of CMs, the model reproduces accurately the final V_T measured after Flash cells have been programmed with different V_{DS} biases. To conclude, although it has been shown that the simulation accuracy can be improved by more complex and physically-based formulae, results suggest that the LEM formula (in some cases corrected by some empirical terms) can be enough for circuit simulations purposes, being a satisfactory tradeoff between accuracy and implementation/computation complexity.

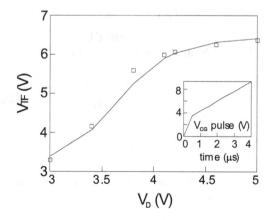

Figure 4-14. Measurements (symbols) and simulations (solid lines) of the final threshold voltage (V_{TF}) after the program of a 0.18 μm Flash memory (L=0.3 μm, W=0.16 μm) with various V_{DS} values (V_{DS}=3-5V).

4.3   CHISEL current modeling

As anticipated in Chapter 2, the CHISEL injection has been introduced in the last years to improve Flash memory program efficiency and overcome the major drawbacks of the classical CHE injection, that are: the large voltage required, the large current involved, and the lack of program control, which imposes the use of Program and Verify algorithms [59-62].

Unfortunately, the physics underneath the CHISEL injection involves mechanisms that are difficult to model using a compact approach, and Monte Carlo transport simulations are usually adopted to simulate CHISEL currents. In fact, only by using such numerical-statistical techniques that are required to model accurately energy distributions of carriers involved in CHISEL injection mechanisms [62-65], CHISEL current can be simulated with a reasonable degree of accuracy.

On the other hand, if an accurate expression for CHISEL carrier energy distribution were derived, some formulae like equations (25)-(27) could be applied to calculate the CHISEL current, thus following the CHE current modeling approach illustrated in Section 4.1.2. Unfortunately, an analytical formula for CHISEL electron energy distribution has not been derived yet, and hardly this task will be solved maintaining a reasonable degree of accuracy. In fact, while for channel hot electrons many approximations in Ref. [53] have been assumed, it appears very complex in CHISEL injection because of the intricate physical mechanisms involved. Such mechanisms, sketched in Figure 4.15 (see also Chapter 2), are briefly described in the following. To generate an electron contributing to the CHISEL current, a channel electron, e1, has to ionize into the drain producing an electron-hole (e2-h2) pair. Then, the so generated hole, h2, heated by the high electric field at the drain junction, has to ionize again to create a new electron-hole pair (e3-h3). Finally, the e3 electron has to gain from the high-field at the drain-body junction enough energy so that when it reaches the Si/SiO_2 interface it can surmount the oxide potential barrier.

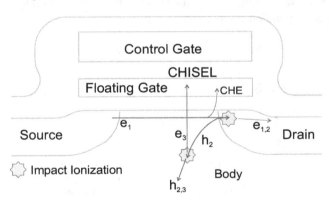

Figure 4-15. Schematic cross section of a Flash memory cell, where physical phenomena involved in CHISEL injection mechanism are evidenced.

As the reader can understand from the above qualitative picture, for an accurate modeling of e3 electron energy distribution it is required not only to reproduce carefully the cascaded sequence of mechanisms concurring to the CHISEL injection, which is a very difficult task considering their inherent complexity and the circuit simulation environment of CMs, but to perform this task adopting a statistical approach, which is needed to reproduce hole and electron energy distributions.

On the other hand, to achieve a first order approximated expression of the CHISEL current, or almost to simulate roughly the dependency of the gate current on the back bias, V_{SB}, CHE formulas proposed in Section 4.1 can be multiplied by an empirical factor exponentially dependent on V_{SB}, modeling the increase of hot electron injection when the body is negatively biased.

REFERENCES

1. A. Kolodny, S. T. K. Nieh, B. Eitan, and J. Shappir, "Analysis and modeling of Floating Gate EEPROM cells," *IEEE Trans. Electron Dev.*, Vol. ED-33(6), pp.835-844, 1986.
2. F. Gigon, "Modeling and simulation of the 16 megabit EPROM cell for write/read operation with a compact SPICE model," *Proc. IEDM*, pp. 205-208, 1990.
3. S. Keeney, F. Piccinini, M. Morelli, A. Mathewson, C. Lombardi, R. Bez, L. Ravazzi, and D. Cantarelli, "Complete transient simulation of flash EEPROM devices," *Proc. IEDM*, pp. 201-204, 1990.
4. J. Suñè, M. Lanzoni, R. Bez, P. Olivo, and B. Riccò, "Transient simulation of the erase cycle of floating gate EEPROMs," *IEDM Tech. Dig.*, pp.905-908, 1991.
5. S. Keeney, R. Bez, D. Cantarelli, F. Piccinini, A. Mathewson, L. Ravazzi, C. Lombardi, "Complete Transient Simulation of Flash EEPROM Devices", *IEEE Trans. Electron Dev.*, Vol. ED-39(12), pp. 2750-2757, 1992.
6. M. Lanzoni, J. Suné, P. Olivo, B. Riccò, "Advanced electrical-level modeling of EEPROM cells", *IEEE Trans. Electron Dev.*, Vol. ED-40(5), pp.951-957, 1993.
7. K.V. Noren and Ming Meng, "Macromodel development for a FLOTOX EEPROM," *IEEE Trans. Electron Dev.*, Vol. ED-45(1), pp.224-229, 1998.
8. M. Lorenzini, R. Vissarion, and M. Rudan, "Three-dimensional modeling of the erasing operation in a submicron Flash-EEPROM memory cell," *IEEE Trans. Electron Dev.*, Vol. ED-46(5), pp.975-983, 1993.
9. S. Chung, C.-M. Yih, S. S. Wu, H. H. Chen, and G. Hong, "A Spice-compatible Flash EEPROM model feasible for transient and program/erase cycling endurance simulation," in *IEDM Tech. Dig.*, pp.179-182, 1999.
10. M. O'Shea, A. Concannon, K. G. McCarthy, B. Lane, A. Mathewson, M. Slotboom, "Macro Model for Flash EEPROM cells," *Proc. ESSDERC*, pp. 352-355, 2000.
11. R. Duane, A. Concannon, P.O'Sullivan, A. Mathewson, "Advanced numerical modelling of non-volatile memory cells", Proc. ESSDERC, pp. 304-307, 1998.
12. R. Bez, D. Cantarelli, and P. Cappelletti, "Experimental transient analysis of the tunnel current in EEPROM cells," *IEEE Trans. Electron Dev.*, Vol. ED-37(4), pp.1081-1086, 1990.
13. L. Esaki, "Long journey into tunneling," *Proc. IEEE*, vol. 62, pp. 825-831, 1974.
14. J. Moll, *Physics of Semiconductors*. New York: McGraw-Hill, 1964.
15. M. Lenziger and E.H. Snow, "Fowler-Nordheim tunneling into thermally grown SiO_2," *J. Appl. Phys.*, vol. 40, no. 1, pp. 278-283, 1969.
16. G. Pananakakis, G. Ghibaudo, and R. Kies, "Temperature dependence of the Fowler-Nordheim current in metal-oxide-degenerate semiconductor structures," *J. Appl. Phys.*, vol. 78, no. 4, pp. 2635-2641, 1995.
17. L. Larcher, A. Paccagnella, and G. Ghidini, " Gate Current in Ultra-thin MOS Capacitors: A New Model of Tunnel Current ", IEEE Trans. Electr. Devices, Vol.48, N.2, pp.271-278, 2001.
18. Z. A. Weinberg, "On tunneling in metal-oxide-silicon structure," *J. Appl. Phys.*, vol. 53, no. 7, pp. 5052-5056, 1982
19. M.V. Fischetti, D.J. Di Maria, L. Dori, J. Batey, E. Tierney, and J. Stasiak, "Ballistic electron transport in thin silicon dioxide films," *Physical Review B*, vol. 35, no. 9, pp. 4404-4415, 1987.
20. O. Briere, K. Barla, A. Halimaoui, and G. Ghibaudo, "Oscillatory behavior of the tunneling current in ultra thin gate dielectrics influence of various physical and technological parameters," *Solid-St. Electron.*, vol. 41, no. 7, pp. 987-990, 1997.

21. J. Maserjian and N. Zamani, "Behavior of the Si/SiO$_2$ interface observed by Fowler-Nordheim tunneling," *J. Appl. Phys.,* vol. 56, no. 1, pp. 559-567, 1982.

22. A. Schenk and G. Heiser, "Modeling and simulation of tunneling through ultra-thin gate dielectrics," *J. Appl. Phys.*, vol. 81, no. 12, pp. 7900-7910, 1997.

23. C. Moglestue, "Self-consistent calculation of electron and hole inversion charges at silicon-silicon dioxide interfaces," *J. Appl. Phys.,* vol. 59, no. 9, pp. 3175-3183, 1986.

24. P. Olivo, J. Sune, and B. Riccò, "Self consistent solution of the Poisson and Schrödinger equations in accumulated semiconductor insulator interface," *J. Appl. Phys.,* vol. 70, no. 1, pp. 337-345, 1991.

25. A. Wettstein, A. Schenk, A. Scholte, and W. Fichtner, "The influnce of localized states on gate tunnel current – Modeling and simulation," in *Technical report No. 97/9 (ETH)*, SISPAD 1997, Boston, September 9, 1997.

26. S.S. Gong, M.E. Burnham, N.D. Theodore, and D.K. Schroeder, "Evaluation of Q$_{BD}$ for electrons tunneling from the Si/SiO$_2$ interface compared to electron tunneling from the poly-Si/SiO$_2$ interface," *IEEE Trans. Electron Dev.*, Vol. ED-40(7), pp.1251-1257, 1993.

27. T. Ando, A. B. Fowler, and F. Stern, "Electronic properties of two-dimensional systems," *Rev. mod. Phys.*, vol. 58, pp. 437-672, 1982.

28. K. S. Krisch, J. D. Bude, and L. Manchanda, "Gate capacitance attenuation in MOS devices with thin gate dielectrics", *IEEE Electron Device Lett.,* vol. 17, no. 11, pp. 521-524, 1996.

29. S. Takagi and A. Toriumi, "Qualitative undestanding of inversion-layer capacitance in Si MOSFET's", *IEEE Trans. Electron Devices,* vol. 42, no. 12, pp. 2125-2130, 1995.

30. N. D. Arora, R. Rios, and D. A. Antoniadis, "Capacitance modeling for deep submicron thin gate oxide MOSFETs", in *Proc. ESSDERC*, pp. 569-572, 1995.

31. J. A. Lòpez-Villanueva, P. Cartujo-Casinello, J. Banqueri, F. Gàmiz, and S. Rodrìguez, "Effects on the inversion layer centroid on MOSFET behavior", *IEEE Trans. Electron Devices,* vol. 44, no. 11, pp. 1915-1922, 1997

32. A. P. Gnädinger and H .E. Talley, "Quantum mechanical calculation of the carrier distribution and the thickness of the inversion layer of a MOS field-effect transistor", *Solid State Electron.*, Vol. 13, pp. 1301-1309, 1970.

33. F. Stern, "Self-consistent results for n-type Si inversion layers", *Physical Review B*, vol. 5, no. 12, pp. 4891-4899, 1972.

34. L. Larcher, P. Pavan, F. Pellizzer, G. Ghidini, "A new model of gate capacitance as a simple tool to extract MOS parameters," *IEEE Trans. Electron Dev.*, Vol. ED-48(4), pp.935-945, 2001.

35. J. Suñè, P. Olivo, and B. Riccò, "Quantum-mechanical modeling of accumulation layers in MOS structure," *IEEE Trans. Electron Devices,* vol. 39, no. 7, pp. 1732-1738, 1992.

36. Y. Tsividis, "Operation and modeling of the MOS transistors", Second Edition, Chap. 2 and 6, McGraw-Hill, 1999.

37. L. Reggiani, "Hot electron transport in semiconductors," Springer Verlag topics in Applied Physics, vol. 58, p. 58, Berlin-Heidelberg, 1985.

38. C. Hu, "Lucky-electron model of channel hot electron emission," in *IEDM Tech. Dig.*, pp.223-226, 1979.

39. S. Tam, P. K. Ko, and C. Hu, "Lucky-electron model of electron injection in MOSFET's," *IEEE Trans. Electron Devices,* Vol. 31, N. 9, pp. 1116-1125, 1984.

40. T. H. Ning, C. M. Osbourn, an H. N. Yu, "Emission probability of hot electrons from silicon into silicon dioxide," *J. Appl. Phys.*, Vol. 48, pp. 286-293, 1977.

41. B. Eitan, and D. Frohman-Bentchkowsky, "Hot-Electron injection into the oxide in n-channel MOS devices," *IEEE Trans. Electron Dev.*, Vol. ED-28(3), pp.328-340, 1981.

42. http://www.semiconductors.philips.com/Philips_Models/

43. 43v3.2. Manual, 1999 U. C. Berkeley, USA, 1999. [available online: http://www-device.eecs.berkeley.edu/~ptm].

44. W. Liu, X. Jin, K. M. Cao, an C. Hu, "44.0.0 MOSFET Model User's Manual," 2000.

45. C. Enz, F. Krummenacher, and E. Vittoz, "An analytical MOS transistor model valid in all regions of operations and dedicated to low voltage and low current applications," Analog Integrated Circuit & Signal Processing, Vol. 8, pp. 83-114, July 1995.

46. M. Bucher, C. Lallament, C. Enz, F. Theodoloz, and F. Krummenacher, "The EPFL-EKV MOSFET model equations for circuit simulation, Version 2.6," Technical Report, Electronics Laboratory, Swiss Federal Institute o Technology (EPFL), June 1997. [available online: http://legwww.epfl.ch/ekv/].

47. G. Gildenblat, N. Arora, R. Sung, and P. Bendix, "Scalable surface potential based compact MOSFET model," Proc. 1997 International Semiconductor Device Research Symposium, p. 33, 1997.
48. T. L. Chen and G. Gildenblat, "Analytical approximation for the MOSFET surface potential," Solid-State Electronics, Vol. 45, pp. 335-339, 2001.
49. http://home.hiroshima-u.ac.jp/usdl/HiSIM.html
50. T. Y. Chan, P. K. Ko, and C. Hu, "Dependence of channel electric field on device scaling," *IEEE Electron Device Lett.*, Vol. 6, N. 10, pp. 551-553, 1985.
51. C. Sodini, P. K. Ko, and J. L. Moll, "The effect of high fields on MOS device and circuit performance," *IEEE Trans. on Electron Devices*, Vol. 31, pp. 1386-1393, 1984.
52. C. Hu, "Hot-electron effects in MOSFET's," *Proc. IEDM*, pp. 176-181, 1983.
53. D. Cassi and B. Riccò, "An analytical model of the energy distribution of hot electrons," IEEE Trans. Electr. Devices, Vol.37, N.6, pp.1514-1521, 1990.
54. C. Fiegna, F. Venturi, M. Melanotte, E. Sangiorgi, and Bruno Riccò, "Simple and efficient modeling of EPROM writing," *IEEE Trans. Electron Dev.*, Vol. ED-38(3), pp.603-610, 1991.
55. K. Hasnat, C.-F. Yeap, S. Jallepalli, W.-K. Shih, S. A. Hareland, V. M. Agostinelli, A. F. Tasch, and C. M. Maziar, "A pseudo-lucky electron model for simulation of electron gate current in submicron nMOSFET's," *IEEE Trans. Electron Dev.*, Vol. ED-43(8), pp. 1264-1273, 1996.
56. K. Hasnat, C.-F. Yeap, S. Jallepalli, S. A. Hareland, W.-K. Shih, V. M. Agostinelli, A. F. Tasch, and C. M. Maziar, "Thermoionic emission model of electron gate current in submicron nMOSFET's," *IEEE Trans. Electron Dev.*, Vol. ED-44(1), pp. 129-138, 1997.
57. Y.A. El Mansy, and A.R. Boothroyd, "A simple-two-dimensional model for IGFET operation in the saturation region," *IEEE Trans. Electron Dev.*, Vol. ED-24, p.254, 1977.
58. P.K. Ko et al., "A unified model for hot-electron currents in MOSFET's," in *IEDM Tech. Dig.*, p.600, 1980
59. B. Riccò, G. Torelli, M. Lanzoni, A. Manstretta, H. E. Maes, D. Montanari, and A. Modelli, "Nonvolatile multilevel memories for digital applications," Proc. of the IEEE, vol. 86, N. 12, pp.2399-2421, 1998.
60. J. D. Bude, A. Frommer, M. R. Pinto, and G. R. Weber, "EEPROM/flash sub 3.0V drain-source bias hot carrier writing," in *IEDM Tech. Dig.*, pp.989-991, 1995.
61. J.D. Bude, M. Mastrapasqua, M.R. Pinto, R.W. Gregor, P.J. Kelley, R.A. Kohler, C.W. Leung, Y. Ma, R.J. McPartland, P.K. Roy, R. Singh, "Secondary Electron Flash – a high performance, low power Flash technology for 0.35 μm and below," in *IEDM Tech. Dig.*, pp.279-282, 1997
62. J.D. Bude, M.R. Pinto, and R.K. Smith, "Monte Carlo simulation of the CHISEL-Flash memory cell," *IEEE Trans. Electron Dev.*, Vol. ED-47(10), pp.1873-1881, 2000.
63. D. Esseni and L. Selmi, "A better understanding of substrate enhanced gate current in VLSI MOSFET's and Flash cells-Part I: Phenomenological Aspects," *IEEE Trans. Electron Dev.*, Vol. ED-46, pp.369-375, 1999.
64. D. Esseni, L. Selmi, A. Ghetti, and E. Sangiorgi, "Injection efficiency of CHISEL gate currents in short MOS devices: physical mechanisms, device implications, and sensitivity to technological parameters," *IEEE Trans. Electron Dev.*, Vol. ED-47(11), pp. 2194-2200, 1999.
65. A. Ghetti, "MOSFET hot-carrier induced gate current simulation by self-consistent silicon/oxide Monte Carlo device simulation," in Proc. Simulation of Semiconductor Processes and Devices (SISPAD), pp. 231-234, 2002.

Chapter 5

FURTHER POSSIBILITIES OF FG DEVICE COMPACT MODELS
Reliability prediction and statistics

The main purpose of this and the next Chapters is to show the new possibilities and application fields of the CM of FG memory devices from a designer's point of view.

As we will show, CMs of FG memories allow not only to reproduce the electrical behavior of FG memory cells in read, program and erase conditions, but also to perform some predictions about reliability (data retention, program and read disturbs) of FG memory devices. Furthermore, accurate models of FG devices permit to gain deep insights into the effects of device parameter statistics, which is strictly related to the whole manufacturing flow, on the electrical performances of FG memories. Moreover the models constitute a powerful tool to optimize program and erase algorithms, and generally to speed-up and simplify the design of the circuitry needed to manage read, program and erase FG memory operations.

Note that the above possibilities and applications illustrate new uses of FG memory CMs that can help both process engineers and circuit designers to develop NV memory product featuring the best tradeoff between reliability and performance.

The chapter is organized as follows. In Paragraph 5.1, it will be described how the charge balance model can be extended to simulate FG memory reliability. Particularly, as done in Chapter 4, one or more voltage controlled current sources modeling the leakage current induced by the aging of the gate oxide will be added to the framework of the model to simulate data retention and program/read disturbs of FG memory devices.

In Paragraph 5.2, it will be shown how FG memory CMs can help designers to make the design phase robust against statistical variations of device parameters. In fact, good CMs must take into account and model correctly the statistical variations of devices' dimensions and performances, thus aiding the designer to cope more effectively with the process induced variability of electrical devices.

1. RELIABILITY PREDICTION

Usually, the reliability of FG memory devices is investigated through experimental techniques and the use of suitable models developed ad-hoc to describe leakage currents through their oxide layers. In fact, leakage currents through gate and interpoly oxides are a serious concern for the reliability of FG memory devices, since they can strongly degrade data retention properties and increase program and read disturbs.

In this scenario, this Paragraph will show that the CM of FG devices (extended to include leakage current effects through voltage controlled current sources) can be a versatile and powerful tool for reliability predictions. In this respect, CMs allow also to bridge the gap between the oxide quality characterization activity performed traditionally on MOS transistors and capacitors, and the actual impact of Stress Induced Leakage Current on FG memory reliability.

Some examples concerning EEPROM memory reliability, and particularly the influence on data retention of Program/Erase (P/E) cycles number and bias conditions, oxide thickness scaling and quality, and storage field, will be illustrated to demonstrate the effectiveness of the CM approach to predict FG memory reliability degradation.

1.1 SILC impact on FG memory reliability

Stress Induced Leakage Current (SILC) is the excess low field current across a thin gate oxide after a high-field stress. This current is a major concern for the reliability of FG memory devices. Moreover, SILC limits strongly the gate oxide thickness reduction, i.e. the FG memory device scaling needed to reduce write voltages and enhance area density [1].

Usually, SILC is measured on large area capacitors [2], even though such measurements are not suited for an accurate evaluation of Flash memory reliability (data retention, program and read disturbs), all statistical aspects related to leakage current variations among different memory cells being completely neglected [3-5].

Since the understanding of the physical mechanisms of oxide degradation and how they affect the FG memory reliability is a key factor for the non-volatile memory technology improvement process, many papers have been proposed in the literature to model and explain the physical mechanism of the SILC conduction, which is generally attributed to the Inelastic Trap Assisted Tunneling (TAT) [6-10].

On the other hand, many efforts have been devoted to the development of new experimental methodologies to directly measure anomalous charge leakage currents in Flash memories [11-13], investigating also their temperature and voltage dependencies [5,14-15]. Other theoretical models describing such anomalous leakage currents and assuming the TAT through one or two traps as conduction mechanism have been proposed [16-17] (in [17], the phonon contribution is also taken into account). Such models have been used also to inspect the spatial distribution of defects generated by high-field stress during Program/Erase (P/E) operations [18-19].

Nevertheless, very few works dealt with the investigation of actual SILC effects on the long-term reliability, such as data retention, endurance, program/read disturbs, and with the tunnel oxide scaling of FG memories [1-2]. In this respect, CMs of FG devices represent a powerful and versatile tool that can be employed to fill this gap and answer to the question: "Which are the actual effects of SILC on FG memory reliability?"

In fact, to perform this task, the model of FG memory devices presented in previous Chapters has to be simply extended by inserting one or more voltage-controlled current sources implementing SILC formula. The current generators will be connected between electrodes comprising the gate oxide portions, which the leakage current flows across.

Thus, the reliability degradation of FG memory device can be effectively predicted by model simulations.

Of course, the accuracy of such predictions depends strictly on the SILC model adopted. To this regard, two different approaches can be followed.

1. Empirical SILC models can be obtained by fitting leakage current measurements, usually performed on large MOS capacitors having the same oxide thickness and the same electrode materials of the FG memory cells considered. Often these models consist of analytical formulae having the minimum number of parameters to facilitate calibration [1,2,20,21]. The advantage of this approach is given by its simplicity and the possibility of correlating fitting parameters to different stress conditions and device manufacturing process recipes. On the other hand, the main drawback is the lack of generality of the SILC formula adopted, which is usually calibrated on devices manufactured in a specific technology, that forces the users to check and eventually modify SILC formula when changing the manufacturing technology.

2. Physically-based SILC models can be adopted to overcome this last disadvantage. In fact, such models describe the physics of SILC conduction mechanisms, and they do not require to be re-targeted at every new technology. Usually, SILC models proposed in the literature assume TAT as the conduction mechanism, and they require the density, the capture cross section and energy depth of defects generated by high-field stress to reproduce experimental data [6-10,16,17,22]. Implementation complexity and computation time that increase significantly compared to empirical models are the price that must be paid to have models that can be applied without changes to FG memory cells with gate oxides manufactured in different technologies and/or using various process recipes.

1.1.1   SILC models proposed in the literature

Various quantitative models of SILC have been proposed in the literature, differing by some features of the underlying physical mechanisms. Early models described SILC in a pragmatic way, by using a modified Fowler-Nordheim expression with a reduced barrier height ($\Phi \approx 0.8\text{-}1$ eV) and a pre-exponential factor depending on bias stress conditions (for example, stress current density and the amount of charge injected during the stress), which acts as fitting parameters [1,20,21]. For these reasons, no insight on the physics of the process was available by using such models.

First SILC models based on a detailed physical description of the conduction mechanism have been proposed in [8,22-24]. Such models assume a multi-phonon assisted tunneling process as conduction mechanism, and compared to models adopting TAT, which appears as the conduction mechanism of choice to model SILC [7,9,10,25-30], their greatest difference is given by the electron coupling to oxide phonons, which results in a series of virtual states in the oxide energy band-gap broadening the trap energy level. Coupling to oxide phonons is represented by an effective phonon energy (~60 meV [24]), and two coupling constants: the lattice relaxation energy and the Huang-Rhys factor.

Despite of minor differences concerning details of physics and/or model implementations, SILC models adopt generally the inelastic Trap Assisted Tunneling as conduction mechanisms that can be viewed as a simple two-step tunneling process. As shown in Fig. 5-1, electrons tunnel from the cathode into a neutral trap: immediately after, the electron-trap center shifts downwards by an energy amount E_P, which has been estimated from quantum-yield experiments and Monte Carlo simulations (E_P=1.2-1.5 eV [7,25]). Subsequently, electrons tunnel out into the anode, and the conduction assisting trap recovers to the original energy level.

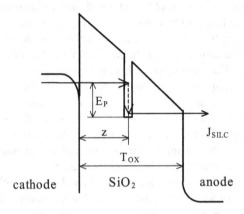

Figure 5-1. Schematic band diagram sketching the SILC physical conduction: the inelastic Trap-Assisted Tunneling. E_P is the energy lost by tunneling electrons.

Several formula differing for minor details of the underlying physics have been proposed in the literature to calculate Stress Induced Leakage Current. Generally, such expressions, that rely to the two-step tunneling model proposed in [31], were derived by balancing cathode-to-trap and trap-to-anode currents through the electron occupancy of conduction assisting traps [7,9,10,29]. Following calculations and assumptions reported in the above referenced paper, that the interested reader is strongly recommended to consult for further details, a quite general expression for the TAT current density (expressed in A/m^3) depending on the trap position, z, and the oxide field can be derived.

$$J_{TAT}(z, F_{OX}) = \sigma \cdot N_T(z) \cdot \frac{J_T(\Phi_B, z, F_{OX}) \cdot J_T(\Phi_B - F_{OX}z + E_P, T_{OX} - z, F_{OX})}{J_T(\Phi_B, z, F_{OX}) + J_T(\Phi_B - F_{OX}z + E_P, T_{OX} - z, F_{OX})} \quad (1)$$

σ and N_T are the capture cross-section and the density of traps. J_T is the uniform tunnel current density through a barrier whose height and thickness are given by Φ_B and z, if the cathode-to-trap tunneling is considered, and by $\Phi_B - F_{OX}\cdot z - E_P$ and $T_{OX}-z$, in the case of trap-to-anode tunneling.

Of course, to evaluate the area density of the TAT current (SILC) flowing at a given F_{OX}, $J_{TAT}(z, F_{OX})$ has to be integrated over the oxide thickness, thus leading to some implementation complexity increase. However, since the last factor of J_{TAT} formula peaks exponentially where cathode-to-trap and the trap-to-anode current densities are equal, which is located approximately in the middle of the oxide for 4-6 nm thick oxides ($z_M \approx T_{OX}/2$), SILC current calculation can be simplified as follows, T_{EFF} being the effective tunneling thickness [9].

$$J_{SILC}(F_{OX}) = \int_0^{T_{OX}} J_{TAT}(z, F_{OX}) dz = \frac{T_{EFF}}{2} \sigma \cdot N_T(z_M) \cdot J_T(\Phi_B, z_M, F_{OX}) \qquad (2)$$

Recently, some changes have been introduced into "classical" SILC models to extend their simulation capabilities. Particularly, some efforts have been devoted to model leakage currents driven by conductive paths comprised of two [16-17] or more traps [32], i.e. to percolation paths. Moreover, the leakage current calculation has been extended also to the case of positively charged traps (whereas traps were generally considered neutral in the most of the papers proposed in the literature on this subject), which requires to account for the modification of the oxide conduction band profile needed for the accurate calculation of the tunnel probability [32-33].

To conclude, it is worth noticing that the low-field excess current can be found not only in electrically stressed oxides, but also in oxides after exposure to ionizing radiation. This Radiation Induced Leakage Current, RILC, displays electrical characteristics that are quite similar to those of SILC [20], and can be modeled in the same way. Thus, by incorporating into the FG memory model a current generator implementing a RILC formula, the reliability degradation occurring in FG memory devices after radiation exposure can be predicted.

1.2 Examples of FG memory device reliability predictions: EEPROM data retention

To demonstrate the effectiveness of CMs to predict FG memory reliability degradation, the influence on EEPROM data retention of P/E cycles, P/E bias conditions, thickness and quality of tunnel oxide, and storage field will be deeply investigated in this Paragraph.

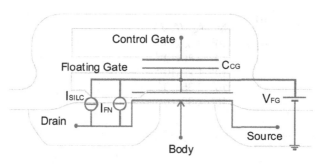

Figure 5-2. The CM of an EEPROM memory cell illustrated in Chapters 3-4 has been extended to simulate SILC-induced EEPROM reliability degradation by including the current generator, I_{SILC}, implementing an analytical SILC formula.

To this purposes, the CM of EEPROM memory cell described in Chapter 4 can be extended to include a voltage-controlled current source implementing the empirical SILC expression proposed in [20], similar also to the one proposed in [1]. As shown in Fig. 5-2, the SILC current generator is connected between the drain and the floating gate since in this region SILC is much larger due to the thinner thickness (~7nm) of tunnel oxide compared to the gate oxide (~20nm).

The empirical formula adopted models the steady-state SILC by using a Fowler-Nordheim-like expression, where the exponential factor is calculated assuming a reduced oxide barrier height ($\Phi \approx 0.8$-1 eV) and the pre-exponential factor, A_{SILC}, depends on the stress current density, J_{STR}, and the amount of charge injected during the electrical stress, Q_{INJ}. SILC transient component is neglected, since it does not lead to a significant FG charge reduction, i.e. to a measurable threshold voltage shift [2].

$$J_{SILC} = A_{SILC} \cdot F_{OX}{}^2 \cdot \exp\left(-\frac{B_{SILC}}{F_{OX}}\right) \tag{3}$$

$$A_{SILC} = C \cdot J_{STR}{}^{\beta} \cdot \exp\left(-D \cdot Q_{INJ}{}^{\alpha}\right) \tag{4}$$

C, D, α, and β are constant parameters, that depend on oxide thickness, stress polarity, and manufacturing technology, and have been derived from gate currents measured on MOS capacitors manufactured using the same process recipe [4,20,21]. To correlate the tunnel oxide degradation induced by high-field stress to the P/E cycles, N_C, and the P/E bias conditions, Q_{INJ} and J_{STR} in Eq. (4) have been replaced by $\Delta Q_{INJ} \cdot N_C$ and the maximum $J_{P/E}$, ΔQ_{INJ} and $J_{P/E}$ being the charge exchanged during a P/E cycle and the P/E current density flowing through the tunnel oxide, respectively [4].

In Fig. 5-3, SILC curves simulated by considering typical P/E conditions (*Program*: D is ramped from 0 to V_R=12 V with a ramp rise time T_R=0.5 ms, CG and B are grounded, and S is left floating; *Erase*: CG is ramped from 0 to V_R=12 V with a ramp rise time T_R=0.5 ms, S, D, and B are grounded) are depicted for different P/E cycles.

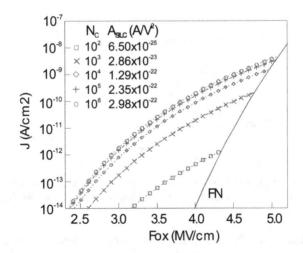

Figure 5-3. SILC curves (symbols) simulated through (1) in a 7 nm thick oxide on increasing the number of P/E cycles (N_C). The A_{SILC} corresponding to N_C is also indicated. The "classic" FN current is shown by a solid line.

Read disturb simulations have demonstrated that SILC is not a concern for EEPROM cells considered, since the oxide field and the time involved in read operations are too low to induce significant FG charge variations, i.e. V_T modification regardless the SILC magnitude [4].

On the contrary, data retention losses are strongly affected by Stress Induced Leakage Current, as predicted by reliability simulations and also confirmed by experimental data.

Fig. 5-4 shows V_T shifts simulated in an erased EEPROM cell left unbiased for ten years at room temperature. In this example, as well as in the following, we have considered only erased EEPROM memory cells, since the oxide field in storage conditions is much larger than in programmed cells, resulting more critical.

There are two aspects of Fig. 5-4 that is worth to stress. First, the threshold voltage reduction occurring after ten years increases with the number of P/E cycles (see dashed lines). This V_T trend is due to the SILC rise on increasing N_C (see Fig. 5-3), whereas the overlap of V_T-time curves for $N_C \leq 10$ is determined by the fact that in these storage field conditions the tunnel current is dominated by the FN component. Second, the ten year threshold voltage after 10^5 cycles does not depend on the initial V_T, i.e. on the initial storage oxide field, $F_{OX,S}$: as shown in Fig. 5-4, V_T-time curves simulated for EEPROM cells after 10^5 P/E cycles assuming different initial V_T (solid lines) converge to a similar value after 2-3 years, which depends on SILC magnitude, but it is independent on the initial V_T, i.e. $F_{OX,S}$.

This first example of reliability simulations shows the powerful prediction capabilities of FG CMs and highlights the importance of a correct experimental methodology for the retention study: in fact, a large initial V_T to accelerate retention extrapolation originates a transient decrease that converges to the "long term" decay curve depending only on the oxide degradation level (N_C).

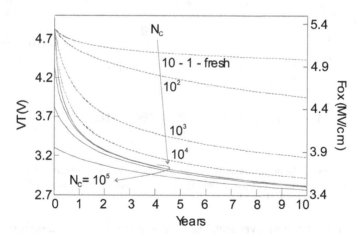

Figure 5-4. V_T decay for a single EEPROM cell left unbiased at room temperature on increasing the cycle number N_C (dashed lines). The oxide field is also indicated. All dashed lines start from the same initial storage field. For $N_C = 10^5$, different initial storage fields have been assumed, and the decay curves are shown by solid lines.

A deeper understanding of the N_C dependence of the threshold voltage shift, which is very important for the data retention survey, can be achieved by using CMs of FG device. In this particular case, the percentage variation of the threshold voltage, $\Delta V_{T(\%)}$, versus N_C for different initial storage fields is shown in Fig. 5-5. Schematically, $\Delta V_{T(\%)}$-N_C curves can be divided in four parts: 1) no increase for $N_C<10$; 2) steep increase for $N_C=10$-10^3; 3) moderate rise for $N_C=10^3$-10^5; 4) saturation for $N_C>10^5$. Interesting insights can be gained from the comparison between $\Delta V_{T(\%)}$ (solid lines) and A_{SILC} (dotted line) curves. Since $\Delta V_{T(\%)}$ is directly proportional to the integral of the leakage current over the time, if we assume that the oxide field remains approximately constant during the retention experiment (this occurs when SILC is too low to induce appreciable FG charge variations), the V_T shift induced by SILC after ten years should be directly proportional to A_{SILC}. As shown in Fig. 5-5, this occurs for $N_C \leq 10^3$, and for this reason this part of the $\Delta V_{T(\%)}$-N_C curve can be called as "A_{SILC} dominated." On the contrary, for $N_C>10^3$, $\Delta V_{T(\%)}$ tends to saturate, whereas A_{SILC} increases. The large A_{SILC} induced by the serious oxide degradation occurring for large N_C enhances significantly the leakage current, lowering the FG voltage. Thus, the oxide field reduces, and the SILC decreases compensating the A_{SILC} increase. Therefore, we can say that this part of $\Delta V_{T(\%)}$-N_C curve is "F_{OX} dominated".

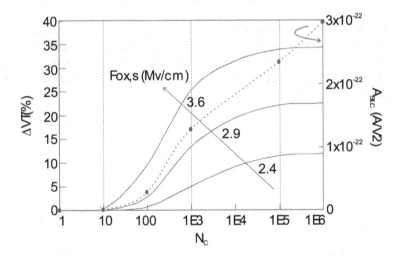

Figure 5-5. Percentage variation of threshold voltage shift (solid lines) simulated in an unbiased 0.35 μm EEPROM cell after ten years, $\Delta V_{T(\%)}$,on increasing N_C for different initial storage field, $F_{OX,S}$. The A_{SILC} trend versus N_C is also shown by dotted lines.

CMs can be used also to investigate the influence of P/E bias conditions (rise time, T_R, and plateau value, V_R, of write voltage ramps) on data retention of FG memories.

Simulations of threshold voltage shifts occurring in unbiased EEPROM cells after 10^5 P/E cycles, that have been programmed and erased adopting different T_R and V_R voltage ramps, are depicted in Fig. 5-6. As shown in this figure, the V_T shift increases on either reducing T_R (faster ramp) or increasing V_R. Furthermore, the rise time of P/E voltage ramps is more influential on the final V_T value than its plateau value. This is due to the fact that program/erase currents, i.e. the damage induced on the tunnel oxide, are larger for faster voltage ramps (see Fig. 4-6), and therefore SILC rises. It is worth noting that what learned by using FG CMs, i.e. that the rate of P/E voltage ramps has a larger influence on EEPROM retention than its plateau value, is very important to design new write conditions. In fact, for P/E voltage ramps of EEPROM devices to be less damaging, they must be as slow as possible, according to memory speed requirements.

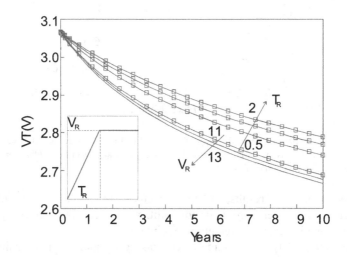

Figure 5-6. V_T decay simulated on an EEPROM cell ($N_C=10^5$) programmed and erased by different T_R and V_R write voltage ramps: symbols refer to $V_R=11V$ and $T_R=$ 0.5, 1, 1.5 and 2 ms voltage ramps; solid lines refer to $T_R=0.5$ ms, $V_R=$ 11, 12 and 13 V voltage ramps.

By using CMs of FG devices, the scaling trends of Flash and EEPROM memory reliability can be reasonably forecasted.

In Fig. 5-7, $\Delta V_{T(\%)}$ is plotted versus the initial storage field ($F_{OX,S}$) for scaled EEPROM cells after 10^5 P/E cycles. EEPROM cells have been scaled according to the constant field scaling rules, i.e. reducing W and L proportionally to T_{OX}. As show in Fig. 5-7, $\Delta V_{T(\%)}$ grows on both increasing the initial storage oxide field and reducing T_{OX} thickness.

To compare data retention properties of scaled EEPROM memory cells, a new quantity, i.e. the initial storage field ensuring $\Delta V_{T(\%)}\leq10\%$ in cells cycled 10^5 times, $F_{OX,T}$, has been defined. $F_{OX,T}$, which has derived from simulations, scales with T_{OX} (see the inset of Fig. 5-7), thus showing that to scale EEPROM memories satisfying the same data retention requirements the storage field has to be reduced in contrast to the constant field scaling rules. Definitely, this demonstrates that the constant field scaling theory cannot be applied to FG devices because of the increase of SILC-related reliability degradation.

Furthermore, the trend of the threshold voltage shift versus T_R and V_R exhibited by the $T_{OX}=7$nm EEPROM has been shown also by scaled EEPROM devices. However, as shown in Fig. 5-8, the V_T shifts are much larger for scaled devices. Particularly, $\Delta V_{T(\%)}$ grows almost linearly on scaling EEPROM devices: a ~1.3% V_T increase for a 0.1 nm reduction of tunnel oxide, confirming again that constant field scaling rules cannot be applied to scale FG devices.

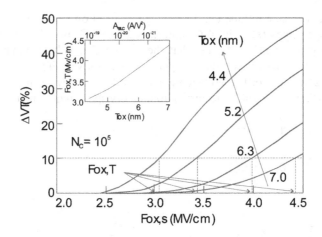

Figure 5-7. $\Delta V_{T(\%)}$ vs. initial storage fields ($F_{OX,S}$) for scaled EEPROM cells cycled 10^5 times. The threshold oxide field, $F_{OX,T}$, is defined as the initial storage field ensuring a $\Delta V_{T(\%)}$ lower than 10% after 10 years, and it is shown in the inset as a function of T_{OX}.

On the basis of these results, i.e. SILC increase in thinner oxides enhances data retention degradation, there are two possibilities to scale FG devices guaranteeing standard Non Volatile Memories retention requirements:

1. to improve the quality of the oxide layer, making it more resistant against electrical stresses, so that the number of defects generated during P/E stresses is reduced;
2. to lower the storage oxide field, so that also the SILC lowers.

Note that while the first point is related to technology improvements, the second one needs to gain the capability to design EEPROM cells with narrower program/erase threshold voltage windows.

Although some statistical data of FG memory reliability issues, for example the dispersion of the V_T distribution, have been completely neglected above, CMs of FG memory devices can be also used to investigate such important aspects.

The dispersion of program and erase V_T distribution are due to several physical reasons related to the statistical nature of device manufacturing process: overall, the most significant factors spreading the V_T distribution are the statistical fluctuations of devices' doping and geometry, that will be analyzed in details in the next paragraph, and the positions of oxide defects. Of course, to be effective from a statistical point of view, CMs have to be able to catch device statistical variations.

The random position of defects within the oxide layer originates a wide range of leakage currents that can differ by several orders of magnitude, and are one of the main physical causes of the V_T distribution spread. In fact, in very small size VLSI devices SILC is related to single defect positions, and it can be strongly and unpredictably enhanced by random build up of charge or defects inside the oxide [34] as well as by the alignment of two or more defects to form a leakage percolation paths through the oxide [16,17,32]. Further, this issue is even more critical when scaling down device sizes: a factor K reduction of device size induces a factor K^3 decrease of the total number of defects inside the oxide, so that the impact of defect fluctuations on the leakage current become more and more important.

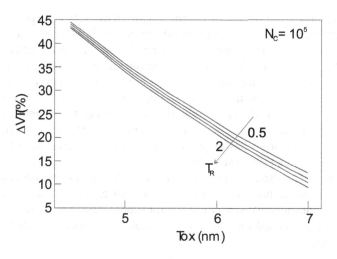

Figure 5-8. $\Delta V_{T(\%)}$ vs. T_{OX} curves simulated in EEPROM memory cells, that have been programmed and erased 10^5 times by using write voltage ramps with the following characteristics: $V_R=12V$ and $T_R=0.5, 1, 1.5,$ and 2 ms.

Of course, to catch the statistical nature of data retention losses in FG memories, leakage current models taking into account the statistics of the defect distribution have to be implemented through some current generators [32]. In this case, CMs will be also a powerful tool to statistically analyze FG device reliability.

2. STATISTICS

Different runs of the same process never give exactly the same output. This simple idea and its statistical consequences have important implications in the modeling and design of integrated circuits.

Each elementary device (resistor, transistor, memory cell, etc...) is different from those used as a reference during the design phase, namely the devices used to extract the model cards, because of random process variation during the device production. Therefore, similarly to the case of supplied voltages and working temperatures, the designer has to cope with a range of possible values for the device parameters, which will span a range of values.

Hence, a robust design must keep all these possible variations into account. Thus, for the design phase to be effective, device CMs taking into account process-induced variations of devices' dimensions and performances must be provided to designers.

In this respect, the model presented in the above Chapters has been designed to fulfill these requirements. In particular, the underlying MOS transistor model reproducing the *dummy cell* allows taking into account variations of parameters as width, length, threshold voltage, gate oxide thickness, etc, while the additional parameters can also easily account for statistical variations due to manufacturing process of FG devices.

Generally, there is a trade-off between designers' requests and specifications on the process variability: on one side designers ask for very tight process controls, whereas on the other side process engineers usually ask for wider process specifications. However, to achieve the best compromise between device performances -including occupied area- and process costs, which are proportional to how tight is the control of the process, it is necessary to guarantee a continuous and deep cooperation between designers and process engineers. The same considerations hold also for Non Volatile Memories' cells: during the design of a new circuit with a new process, an agreement on the specifications of the memory cells' performances has to be found including the variations, and the design will have to take into account such variability.

The specifications usually concern a subset of the process parameters: this will be called the set of the key parameters; and let's call "process range" the set of all the values of all the key parameters that satisfy the specifications. Many approaches can be followed to cope with the mentioned variability, once agreed on the specifications for the different process parameters.

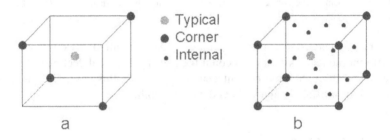

a   b

Figure 5-9. Exemplification in 3 dimensions of (a) the worst case approach with 4 corner models (b) approach spanning the complete process range.

A first, very popular, way to account for process variations into the design follows this idea: if the process is within its "process space" the device is expected to work properly. In the frame of this approach, where a one-to-one relation between dice within specifications and functional dice is expected, the commonly used "worst cases" design is a satisfactory methodology. It consists of simulating, besides the typical case, a properly selected subset of the corners of the process range hyper-parallelepiped (see Fig. 5-9(a)). The minimum number of points considered is 2, usually selected in such a way that the current is maximized and the delay time is minimized, for the first point, while for the second one the current is minimized and the delay time is maximized. The maximum number will be 2^n, where n is the number of key parameters: one simulation point for each of the vertexes of the hyper-parallelepiped. Still within the same approach, to increase the confidence on the results, a set of simulations spanning the complete "process space" can be run (see Fig. 5-9(b)). Clearly this increases the simulation time.

A different approach consists in using the probability density function and the correlation between the key parameters. This approach allows to neglect points in the process range that will be never, or very unlikely, reached. In fact, if parameters are highly correlated, some points can be neglected (points B in Fig. 5-10). This approach, which is theoretically very valuable, could be used to find an optimum solution to the trade off between occupied area and performances on one side and yield on the other. Unfortunately, on the practical side, this approach is very costly as it is very difficult and long to collect all the process parameters distributions and their correlations. Finally, it is worth noticing that this approach can be applied to a full device but also to particularly critical blocks.

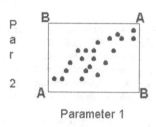

Figure 5-10. Example in 2 dimensions of points (B) in the "process range" that will never be reached.

REFERENCES

1. Jan de Blauwe, Jan Van Houdt, Dirk Wellekens, Guido Groseneken, and Herman E. Maes, "SILC-related effects in flash-EEPROM's-Part I: A quantitative model for steady-state SILC", *IEEE Trans. Electron Devices*, vol. 45, no. 8, pp. 1751-1760, 1998.
2. Jan de Blauwe, Jan Van Houdt, Dirk Wellekens, Guido Groseneken, and Herman E. Maes, "SILC-related effects in flash-EEPROM's-Part II: Prediction of steady-state SILC-related disturb characteristic," *IEEE Trans. Electron Devices*, vol. 45, no. 8, pp. 1751-1760, 1998.
3. S. Satoh, G. Hemink, K. Hatakeyama, and S. Aritome, "Stress-Induced Leakage Current of tunnel oxide derived from Flash memory read-disturb characteristics," *IEEE Trans. Electron Dev.*, Vol. 45, N. 2, pp. 482-486, 1998.
4. L. Larcher, S. Bertulu, P. Pavan, "SILC effects on EEPROM memory cell reliability," *IEEE Trans. Device and Material Reliability*, Vol. 2, N. 1, pp. 13-18, 2002.
5. H.P. Belgal, N. Righos, I. Kalastirsky, J.J. Peterson, R. Shiner, and N. Mielke, "A new reliability model for post-cycling charge retention in Flash memories," in *Proc. 40th IEEE-IRPS*, Dallas (USA), pp.7-20, 2002.
6. R. Moazzami and C. Hu, "Stess-induced current in thin silicon dioxide films," in *Proc. IEDM*, San Francisco (USA), pp. 139-142, 1992.
7. L. Larcher, A. Paccagnella, and G. Ghidini, "A new model of Stress Induced Leakage Current in gate oxides", *IEEE Trans. Electron Dev.*, Vol. 48, N. 2, pp. 285-288, 2001.
8. Kiyohiko Sakakibara, Natsuo Ajika, Katsumi Eikyu, Kiyoshi Ishikawa, and Hirokazu Miyoshi, "A quantitative analysis of time-decay reproducible stress-induced leakage current in SiO₂ films", *IEEE Trans. Electron Devices*, vol. 44, no. 6, pp. 1002-1007, 1997
9. Shin-ichi Takagi, Naoki Yasuda, and Akira Toriumi, "A new I-V model for stress-induced leakage current including inelastic tunneling", *IEEE Trans. Electron Devices*, vol. 46, no. 2, pp. 348-354, 1999.
10. B. Riccò, G. Gozzi, and M. Lanzoni, "Modeling and simulation of Stress-Induced Leakage Current in ultrathin SiO₂ films," *IEEE Trans. Electron Devices*, vol. 45, no. 7, pp. 1554-1560, 1998.

11. C. Lam, T. Sunaga, Y. Igarashi, M. Ichinose, K. Kitamura, C. Willets, J. Johnson, S. Mittl, F. White, H. Tang, T.-C. Chen, "Anomalous low temperature charge leakage mechanism in ULSI Flash memories," in *Proc. IEDM*, San Francisco (USA), pp. 335-338, 2000.

12. H. Kameyama, Y. Okuyama, S. Kamohara, K. Kubota, H. Kume, K. Okuyama, Y. Manabe, A. Nozoe, H. Uchida, M. Hidaka, and K. Ogura, "A new data retention mechanism after endurance stress on Flash memory," in *Proc. 38th IEEE-IRPS*, San jose (USA), pp. 194-199, 2000.

13. D. Ielmini, A. S. Spinelli, A. L. Lacaita, and A. Modelli, "Equivalent cell approach for extraction of the SILC distribution in Flash EEPROM cells," *IEEE Electron Devices Letters*, Vol. 23, N. 1, pp. 40-42, 2002.

14. P. J. Kuhn, A. Hoefler, T. Harp, B. Hornung, "A reliability methodology for low temperature data retention in floating gate non-volatile memories," in *Proc. 39th IEEE-IRPS*, Orlando (USA), pp. 266-270, 2001.

15. A. Modelli, F. Gilardoni, D. Ielmini, and A. S. Spinelli, "A new conduction mechanism for the anomalous cells in thin oxide Flash EEPROMs," in *Proc. 39th IEEE-IRPS*, Orlando (USA), pp. 61-66, 2001.

16. D. Ielmini, A. S. Spinelli, A. L. Lacaita, and A. Modelli, "A new two-trap tunneling model for the anomalous SILC in Flash memories," in Proc. *INFOS*, pp. 39-40, 2001.

17. F. Schuler, R. Degreave, P. Hendrickx, D. Wellekens, „Physical description of anoumaluos charge loss in floating gate based NVM's and identification of its dominant parameter," in *Proc. 40th IEEE-IRPS*, Dallas (USA), pp. 26-33, 2002.

18. D. Ielmini, A. S. Spinelli, A. L. Lacaita, and A. Modelli, "Statistical modeling or reliability and scaling projections for Flash memories," in *Proc. IEDM*, Washington DC (USA), pp. 703-706, 2001.

19. R. Degraeve, F. Schuler, M. Lorenzini, D. Wellekens, P. Hendrickx, J. Van Houdt, L. Haspeslagh, G. Groseneken, G. Tempel, "Analitycal model for failure rate prediction due to anomalous charge loss of Flash memories," in *Proc. IEDM*, Washington DC (USA), pp. 699-702, 2001.

20. A. Scarpa, P. Reis, G. Ghibaudo, A. Paccagnella, G. Pananakakis, J. Brini, G. Ghidini and C. Papadas, "Stress Induced Leakage Current Dependence on Oxide Thickness, Technology and stress level," in Proc. of the 27th European Solid-State Device Research Conference (ESSDERC 97), pp. 592-595, Stuttgart (Germany), 22-24 September 1997.

21. A. Scarpa, *Ph.D. Thesis,* Chap.7, University of Padova, Italy, 1998.

22. K. Sakakibara, N. Ajika, M. Hatanaka, and H. Miyoshi, "A quantitative analysis of stress induced excess current in SiO_2 films", in *IRPS Tech. Dig.*, p. 100, 1996

23. A. Schenk and H. Hermann, "A new model for long term charge loss in EPROMs," *Ext. Abstract of the International Conference on Solid State Devices and Materials (SSDM)*, pp. 494-496, Yokohama (Japan), 1994.

24. M. Hermann and A. Schenk, „Field and high temperature dependence on the long term charge loss in erasable programmable read only memories: Measurements and Modeling," *J. Appl. Phys.*, Vol. 77, N. 9, pp. 4522-4540, 1995.

25. Shin-ichi Takagi, Naoki Yasuda and Akira Toriumi, "Experimental evidence of inelastic tunneling and new I-V model for stress-induced leakage current", in *IEDM Tech. Dig.*, pp. 323-326, 1996.

26. J. Wu, L.F. Register, and E. Rosenbaum, "Trap-assisted tunneling current through ultra-thin oxide", in *IRPS Tech. Dig.*, pp. 396-399, 1999.

27. Paul E. Nicollian, Mark Rodder, Douglas T. Grider, Peijun Chen, Robert M. Wallace, and Sunil V. Hattangady, "Low voltage stress-induced leakage current in ultrathin gate oxides", in *IRPS Tech. Dig.*, pp. 400-404, 1999.

28. C.T. Liu, A. Ghetti, Y. Ma, G. Alers, C.P. Chang, K.P. Cheung, J.I. Colonell, W.Y.C. Lai, C.S. Pai, R. Liu, H. Vaidya, and J.T. Clemens, "Intrinsic and stress-induced traps in the direct tunneling current of 2.3 – 3.8nm oxides and unified characterization methodologies of sub-3nm oxides", in *IEDM Tech. Dig.*, pp. 85-88, 1997.

29. D. Ielmini, A. S. Spinelli, M. A. Rigamonti, A. L. Lacaita, "Modeling of SILC based on electron and hole tunneling. II. Steady-state," *IEEE Trans. Electron Dev.*, Vol. 47, N. 6, pp. 1266-1272, 2000.

30. Anthony I. Chou, Kafai Lai, Kiran Kumar, Prasenjit Chowdhury, and Jack C. Lee, "Modeling of stress-induced leakage current in ultrathin oxides with the trap-assisted tunneling mechanism", *Appl. Phys. Lett.*, vol. 70, no. 25, pp. 3407-3409, 1997.

31.N. Yasuda, N. Patel, and A. Toriumi, "A two-step tunnelling model for stress induced leakage currents in ultra-thin silicon dioside films," *Ext. Abst. Solid State Devices and Materials*, pp. 847-850, 1993.

32.L. Larcher, "Statistical Simulation of Leakage Currents in MOS and Flash Memory Devices with a new multi-Phonon Trap-Assisted Tunneling Model," to be published on Trans. On Electron Devices.

33.G. Cellere, L. Larcher, M.G. Valentini, and A. Paccagnella, "Micro breakdown in small-area ultra-thin gate oxide," on Trans. on Electron Devices, Vol.49, N.8, pp.1367-1374, 2002.

34.P. Cappelletti, R. Bez, D. Cantarelli, and L. Frattin, "Failure mechanism of Flash cell in program/erase cycling," in *IEDM Tech. Dig.*, 1994, pp. 291-294.

Chapter 6

NON VOLATILE MEMORY DEVICES
Architecture and Operating Modes

G. Campardo, R. Micheloni
STMicroelectronics, Memory Product Group – Flash Division, Agrate Brianza (MI), Italy
giovanni.campardo@st.com, rino.micheloni@st.com

In the previous chapters, the mechanisms that govern the behavior of a floating gate cell have been investigated; the aim of this last chapter is to describe how these memory cells are used, with a special emphasis on Flash products. We will show how the memory cells are connected in a commercial device to form the matrix, together with their connection to the circuitry that allows a user to program, erase and read the memory content.

The analysis of the various topics is performed based on the elementary blocks composing the whole device, to give a sufficiently precise and exhaustive idea of both the structure and the required functionality of a memory and its main parts. The topics can be thoroughly studied exploiting the vast existing bibliography, where details related to the circuital issues can be found.

To achieve the desired result, several figures have been drawn to show the relative position of the various blocks and their connections. Finally, the device under analysis should be considered as an example since, for sake of simplicity and clarity of description, it represents a real device where some blocks have been either added or removed.

After reading this Chapter, it will be clear that a Compact Model of a Floating Gate device is needed, since it is central in many building blocks constituting a real Non-Volatile Memory device.

1. BASIC ELEMENTS

Figure 6-1 shows the photo of a cross section of a matrix column (bit line) where the basic elements composing a Flash cell can be seen: the *control gate*, obtained by overlapping two layers, the lower being the polysilicon of the control gate (poly2) and the upper being the silicide to reduce row resistivity; the *interpoly oxide*, realized with several layers to improve its retention quality; the *floating gate*, called poly1 to distinguish it from poly2; and the *tunnel oxide*. Furthermore, both source and drain junctions are shown, as well as bit line and the drain contacts of the cell.

Both drain and source contacts are shared among adjacent cells, in order to optimize area use inside the matrix. This kind of configuration is known as *NOR* type because of the way the cells are connected between ground (at the source) and the bit line (at the drain). Other architectures exist but the focus of this chapter is on NOR type Flash memories. Figure 6-1 also concisely shows the biasing used in the different operating modes.

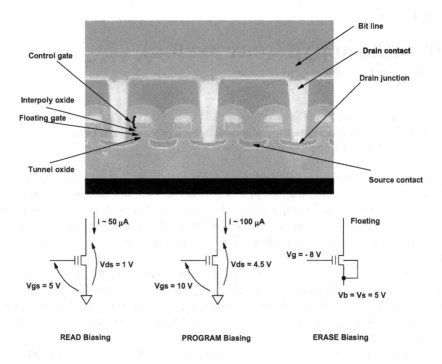

Figure 6-1. Flash cell cross section and biasing voltages used during read, program and erase phases.

1.1   Read biasing

The read operation senses the current flowing through the Flash cell itself. Both program and erase modify the threshold voltage of the memory cell. Usually, programmed cells are those featuring an excess of electrons trapped in the floating gate and therefore a high threshold voltage, while the erased ones have a low threshold voltage. Consequently, applied voltages being equal, a programmed cell sinks less current than an erased one. Read operation consists in biasing the cells and sensing the sunk current by evaluating the voltage drop produced by the current itself onto a resistive load.

The circuit responsible for this operation is known as *sense amplifier* and it is composed, in principle, by a current-to-voltage converter and a voltage comparator. This last one compares the voltage resulting from cell's current conversion with a reference voltage that is usually obtained by reading a cell whose current is well known (reference cell). The precision of this kind of read is in the order of some microampere, and the time required to perform the operation is some tens of nanoseconds.

In the past, this operation was simulated using as a cell model a MOS transistor model, where the threshold voltage was adjusted "by-hand" to simulate the programmed or the erased state of the memory cell. The new CM proposed in this book has been successfully adopted in the design flow of sense amplifier to improve the simulation accuracy: in fact, this model provides simulations predictions closer to the actual circuit performance without increase of simulation time.

When dealing with read, the designer pay a special attention not only to execution speed and precision, which takes with it all the issues related to parasitic components (resistance towards the various terminals, unwanted voltage drops on power supplies, parasitic capacitances, etc.), but also to the precision of the voltages used. It is important as well to avoid any electrical stress on those un-addressed cells that share either the selected bit line or word line. In particular, the value of the drain voltage to be used for read is chosen small enough to prevent spurious programming and at the same time, since it is a current-read, high enough to allow a current flow that can be measured in the desired time.

1.2   Program biasing

In this situation, the phenomenon known as Channel Hot Electrons (CHE) is used to inject negative charge into the floating gate in order to modify the threshold of the cell. To achieve this result, a current must flow into the channel of the cell, applying a high voltage to both the drain and the control gate; values for today's Flash processes are around 4.5 V and 10 V respectively.

The current flowing into the channel under these assumptions is in the range of 100 μA.

Both the voltages and currents required for a proper operation of a Flash memory have varied over the years mainly because of the need of decreasing power supply voltages. Older devices had a 5 V power supply (V_{DD}) together with a separate pin for a 12 V supply (V_{PP}) used for program and erase, then voltages has become lower and lower. Nowadays, a single power supply as low as 1.65 V is used, and 0.9 V has been a realistic target for a while! The wide spreading of portable devices like the mobile phone, digital still cameras, MP3 players etc. has caused a boost towards a dramatic reduction of power consumption requirements to prolong battery life. Among the other consequences, programming current of the cells must be reduced: indeed, it passed from 1 mA to 100 μA and even lower.

The main limitation of CHE program is that a current must flow in the channel of the cell: since several cells are programmed in parallel to reduce program time (today up to 64 cells are programmed at the same time), on-chip charge pumps are required to generate the required current, and the impact on area occupation is not negligible at all. The injection of electrons into the floating gate can also take place by means of the Fowler-Nordheim tunneling effect. In this case, consumption of the cell is virtually reduced to zero, but programming time is increased from some microseconds to some milliseconds. NAND-type architecture is based on this mechanism. Greatest care when designing a program circuit is paid to the correct generation of all the voltages for the different nodes at the proper time. Furthermore it is very important not to expose the cells to electrical stresses that might alter the stored information. Last but not least it is essential to verify the status of every single cell immediately after the programming pulse.

Also in this case, a simple CM of the FG cell including a correct simulation of CHE and FN mechanisms is desired to optimize program/verify cycles, voltage generators, etc., and to evaluate power consumption.

1.3   Erase biasing

The possibility of being electrically erased has given a great appeal to the Flash cell from a market perspective. Until some years ago, EPROM (Erasable Programmable Read Only Memory) has been the most widely used memory device. EPROM is programmed by CHE and it is erased by UV radiation exposure: after about 20 minutes, all the floating gates revert their state to the logic "1", i.e. electrical neutrality. The drawback is that erase implies the removal of the device from the board to be placed under the UV lamp; therefore the usage of EPROM has some limitations, but on the other hand these devices are simpler, since the erase operation takes place on the entire matrix at the same time. As already explained in the introductive chapters, another type of non-volatile memories exists, the EEPROM (Electrically Erasable Programmable Read Only Memory), where Fowler-Nordheim tunneling effect is used to program and erase the cells on a byte-by-byte basis. This is achieved thanks to the presence of a selection transistor for each cell, at the cost of very inefficient area utilization: just to make a comparison, today's maximum density of an EEPROM is 1 Mb, against 256 Mb of a Flash.

The Flash memory combines CHE programming of the EPROM with FN tunneling with modified erase granularity. The smallest set of cells that can be erased in parallel is called sector. A typical size of a sector is 1 Mb, a good trade-off between the user's needs and the size of the devices that are sensible to both the number and the size of the sectors. In order to allow independent erase, each sector must be separated from the others to avoid any interaction and, therefore, undesired stress.

The main issue for a Flash device is related to the number of erase-program cycles, whose limit is usually fixed at 100,000 cycles for each sector. Of course, erase and programming operations cause stress onto the cells resulting in aging of the oxides and performance degradation, above all under a charge retention point of view. Also in this case, reliability predictions that can be obtained with the CM proposed in this book can be used to correctly design ancillary circuits in EEPROM and Flash memory cells.

Erase strategy for Flash devices has been developed in parallel with the applications; as is the case of program, low voltage requirements imposed modifications to the memory organization, also modifying erase mode.

First generation Flash memories inherited from the EPROM the external V_{PP} pin, which was put at 12 V during erase. In this case the gate of the cells of the sector under erase was set to ground, together with the body, and 12 V were applied to the source. Erase is assisted by a current due to Band-to-Band tunneling. The value of this parasitic current is in the order of $1 \div 10$ nA per cell, therefore it is about $1 \div 10$ mA for a 1 Mb sector. The need of scaling down supply voltages and therefore of generating erase voltages using internal charge pumps, intrinsically current-limited, has led to the partition of the voltages between gate and source and, then, to the introduction of the triple well. Triple well CMOS process allowed for complete elimination of the spurious current by means of the configuration shown in Figure 6-1. Drain node is left floating during erase to avoid any voltage drop (and current consumption) that would make the applied voltages useless.

2. MAIN BUILDING BLOCKS OF THE DEVICE

Figure 6-2 shows a photo of a Flash memory device where the main blocks are highlighted. The represented device is a 64 Mb fabricated in CMOS 0.18μm technology (2 levels of polysilicon and 3 levels of metal). Overall size is about 40 mm^2.

Figure 6-3 shows, for sake of simplicity, only the circuital blocks previously highlighted that we are going to detail in the following.

Let's start from the sectors: there are 64 of them, of 1 Mb each, and they are surrounded by blocks called "Local Row Decoder" and "Local Column Decoder". To obtain the physical separation of the sector, in order to avoid any cross-interference during the different operating phases of the memory, a hierarchical decoding technique has been implemented. Addressing of the single cell, or to be more precise of the single byte (8 cells simultaneously) or word (16 cells simultaneously), or even double word (32 cells simultaneously) is achieved by a circuitry that select both the corresponding rows and columns to read or program the cells belonging to the same byte. The row/column crossing identifies the addressed cell.

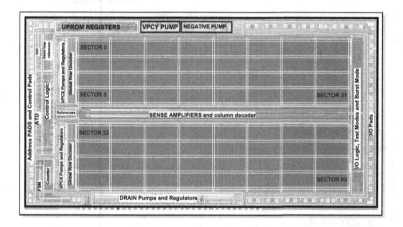

Figure 6-2. Photo of a 64Mb chip, with the main circuital blocks superimposed.

To avoid useless electrical stress, some "switches" are used, i.e. the local decoders, which are enabled to pass the required voltages to the addressed sector only, thus biasing only the cells inside such sector. In the example shown, the 64 sectors are separated each other by these hierarchical decoders with enormous advantages in terms of cell reliability during cycling.

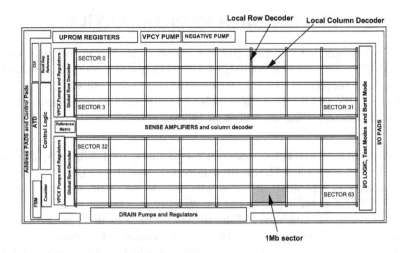

Figure 6-3. The blocks of Figure 6-2 with a sector and hierarchical decoding highlighted.

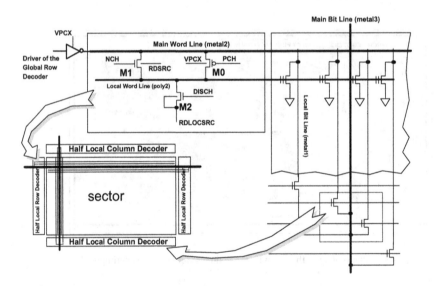

Figure 6-4. A sector of Figure 6-2 where both row and column hierarchical decoders are highlighted.

Figure 6-4 depicts the circuital organization of the hierarchical decoders of a sector. Local column decoding is composed of 4 pass transistors that are connected to the same column, identified in the picture as Main Bit Line. These pass transistors transfer the voltages from the Main to the Local Bit Line. Local Bit Line is realized in metal1, while the Global Bit Line is in metal3. Column decoder is divided into two parts, one placed above and the other below the sector, in order to easily accommodate the pitch of the cell: the transistors contained in the box shown by the arrow are placed in the lower local decoder, while the external ones are placed above the sector.

Local row decoding is instead more complex, since several voltages must be transferred. The selected row requires a positive voltage during read and program, and a negative one in erase, while the unselected rows must be biased to ground in any case (this is true for all the unselected rows of the addressed sector as well as for those of the un-addressed sectors). All these features are implemented thanks to a triple switch as shown in the picture.

In this case the Main Word Line is realized in metal2, while the Local Word Line is in poly2. The choice of assigning either metal2 or metal3 to the Main Word Line rather than to the Main Bit Line depends on the technological rules that set the allowable sizes of the different layers.

It is convenient to cut in two parts the local row decoders in the same way as it has been done for the local column decoders, one on the left and the other on the right of the sector. Local row decoders are driven by global row decoders, represented in the figure as an inverter supplied by the output of a properly filtered charge pump (VPCX). In the example, global row decoding is placed to the left of the matrix and it is divided into two decoders that respectively address the sectors placed above (0 to 31) and those placed below (32 to 63) the central channel.

The central channel is occupied by the global column decoders, whose task is to collect the Global Bit Lines and to connect them to the sense amplifiers. In our case, the overall width of output data is 32 bit, i.e. a double word; therefore the central channel should host at least 32 sense amplifiers, since the read is performed in parallel on all the cells composing the double word itself.

The sense amplifier is the heart of the device, since the most important parameter when evaluating a memory, the access time (defined as the time required to have on the output pads the data stored in the cells whose address is applied to the input pads) is intimately related to the performance of this circuit. A great improvement to the design of the sense amplifier is achieved by using an accurate FG memory cell model including reliable worst cases or, even better, statistical aspects.

Left to the central channel there is a block called "Reference Matrix": it is composed of a small matrix containing the cells used as a reference during the various phases. During read, as we have seen, the current of a cell is used as a term of comparison against the current sunk by the addressed cell to sense a logic value of either "1" or "0". To evaluate the effect of a program operation, the current sunk by the addressed cell is compared with that of a cell (Program Verify) whose threshold has been fixed to the minimum allowable threshold value for a programmed cell. The same is done for a cell under erase, where the reference cell (Erase Verify) has a threshold whose value is the maximum allowable for an erased cell.

Finally, the electrical erase operation not only removes the surplus negative charges from the floating gate, but it might also generate a surplus of positive charges. Such a cell is called depleted since its threshold is negative and it can sink current even if it is not addressed. Therefore these cells can induce errors during read if they are on the same bit line as the programmed cells. The read of a programmed cell implies that the current flowing through the cell is lower than that of the reference but, if a depleted cell is present, the current of this latter sums up to the current of the programmed cell. So the overall current might erroneously result greater than the reference one, causing a read error.

To solve this issue, the erase is followed by a check, automatically performed on all the cells of the sector, to find out the depleted cells and to move them in the correct positive threshold distribution by means of a controlled programming. The search for these cells is done using a cell with a proper threshold called Depletion Verify.

All these cells are placed in the reference matrix and their thresholds are appropriately set during the factory tests.

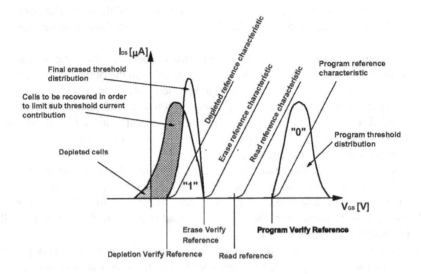

Figure 6-5. Summary of the cells inside the Reference Matrix and their relative position to the "1" and "0" distributions.

This description helps also to understand how strategic can be the correct CM of a single cell, not only for the correct functioning simulation, but also for reliability predictions. For example, one can simulate the impact of the degradation of the cell in the reference matrix on the performance of the device.

Figure 6-5 summarizes what we said about the reference cells and their position relatively to the distributions for the erased and programmed matrix cells.

Next to the decoders, the charge pumps supplying row decoders in both read and program are placed. These charge pumps, like all those currently used in the Flash memories, are realized with the diode-capacitor technique. The output of the pump is regulated by a feedback loop controlled circuit: the generated voltage is called VPCX and it takes the positive values that are used to read and program.

To the leftmost side (Fig. 6-3), the input pads are present, i.e. the addresses and the control pads that, in the case of a memory, are mainly CE# (Chip Enable bar) whose task is to place the device in standby, that is an operating mode featuring a consumption of few microampere; OE# (Output Enable bar) whose task is to place the output buffers in high impedance allowing to the other devices present on the board to drive the output bus; WE# (Write Enable bar) that drives the program and the command phases to the device[1].

---

[1] Control signals are active low because of a legacy from NMOS devices, whose input buffers do not dissipate if the input value is low, since the pull down is turned off.

Moving from left to right the ATD block (Address Transition Detector) is found, in a central position. To explain the operation of this block it is necessary to make a digression on the nature of a Flash memory.

Two main categories of devices exist: synchronous, i.e. whose functions are governed by an external clock, and asynchronous, i.e. whose functions do not obey to a clock.

The former execute their operations sequentially, using the external clock to time the different phases of the tasks. A good example is the microprocessor that executes in this way the instructions that constitute the program. The main drawback of this kind of device is that the single operations, that might take place in a time smaller that the assigned clock cycle, cannot be used to speed up the system.

On the other hand, asynchronous systems do not work according to a clock; every task completes in the time it requires and the result is always achieved at its maximum speed. The issue in this case is that the single operations must be designed to prevent any overlapping due to process, temperature and supply variations that might invalidate the final result.

Memories, at least the non-volatile ones, belong to the categories of the asynchronous devices. Anyway it is really useful to have a signal of synchronism that is asserted any time a new read phase starts. Therefore an ATD circuitry is designed, which is sensitive to address variations[2] and which produces a pulse that acts as a pseudo-clock, triggering all the operations required for a read: pre-charge of the nodes of the sense amplifier, recharge of the boosted nodes [3], opening of the latches towards the output buffers etc.

The use of this technique is very useful since it allows to temporally divide the different operations and therefore to have a reciprocal influence as smaller as possible. The drawback is that timing windows must be guaranteed to execute the operations, so appropriate margins must be taken, thus increasing the overall access time.

Above the ATD block, the CUI (Command User Interface) is found. Task of this block, designed either as a sequencer or as a finite state machine, is to interpret the commands provided by the user to the device. In the command set, we can find: program, erase of a specific sector, read of the status register (which allows monitoring the status of the device during the various phases), program or erase suspend etc. The possibility of suspending a modify operation lets the user access the memory content at any time; without this feature, the system would "lose" the control of the memory for several microseconds (in case of program) or for an entire second (in case of erase).

Below the ATD block, the FSM (Finite State Machine) is found. Task of this block is to execute the program and erase algorithm, working as a small microprocessor able to perform comparisons, conditional branches and counts. This last function, performed by the Counter block, is vital to control the length and the number of both erase and program pulses; each program and erase is indeed performed using a predefined sequence of pulses and subsequent verifications.

Next to the CUI, the Band Gap Reference is found: this circuit provides a voltage that is stable with respect to both temperature and supply voltage variations, and it is used in the state-of-the-art non-volatile memories to control the voltages during both program and erase phases. EPROM memories allowed a variation of the drain voltage around its

---

[2] In case of read from address, the trigger is given by any address variation, in case of read from CE# the trigger is provided by the edge of the CE#.

[3] Boosting is the operation by which a node, initially floating, is brought to a voltage whose value is higher, in modulo, than power supply.

typical value of some hundreds of millivolt. Today, cells' size requires a greater precision for the voltages, especially the drain voltage during program and gate voltage during verify operations. To solve this issue, the circuit known as *band-gap-reference* is used, which utilizes the possibility of compensating temperature variation of the V_{BE} of a bipolar transistor with the variation of a voltage drop that is proportional to the thermal potential ($V_t =KT/q$), where K is Boltzmann's constant, q is the charge of the electron and T is the absolute temperature.

Let's move to the blocks in the upper part of the device (Fig. 6-3).

Starting from the left, the UPROM block can be seen. Fabrication of a memory device containing millions of cells implies the issue of reproducing it despite defectivity. In other words, the probability that one or more memory cells are faulty or anyway do not work properly is very high, because fabrication process are very complex and several parameters must be controlled. All the memories contain more cells than those normally addressable; these cells, called redundancy cells, are used to "substitute" defective cells (if any) detected during the test phases following the fabrication of the device. There are non-volatile registers composed of other non-volatile memory cells, called UPROM, whose task is to store the addresses of the faulty bits.

Any time an address is provided to the memory, it is compared against the content of the UPROM. In case of match (therefore the address corresponds to faulty locations) the normal decoders are turned off and another specific one is enabled, which activates the redundancy rows and/or columns. Then an address transformation occurs, and a new memory location, outside the externally visible addressable space, is accessed.

Next to the UPROM, the VPCY block, used to supply the column decoders during program, can be found. In fact, to avoid an excessive voltage drop on the pass transistors of the decoding, their driving voltage is improved instead of increasing too much their size.

The last block of the upper part is the NEGATIVE PUMP, which is composed by the charge pumps that generate the negative voltages used by the row decoders during erase. This block can also be used to supply the negative voltage to the body of the sector during program to improve its efficiency, exploiting CHISEL mechanism.

In the lower side of the device, the DRAIN Pumps and Regulators: block can be found: it is the charge pump that provides the positive voltage used to bias the drain of the cells during program and to bias the matrix body, shorted with the source, during erase. A lower output voltage and a much higher capability of providing current characterize this charge pump, with respect to the previously described VPCX. Furthermore, the required precision is different for the two terminals of the cell. It is then reasonable that the two pumps are designed separately, in order to better optimize their performances.

Let's analyze the two blocks to the rightmost of the device (Fig. 6-3).

The first, called I/O LOGIC, Test Modes and Burst Mode, carries out different tasks that, for sake of simplicity, have been merged.

The first of these functions is the managing of the I/O logic that receives the data from the sense amplifiers and distributes them to the output buffers. These data are stored in volatile registers and then sent to the output buffers. In the most advanced memories, data processing logic to be performed before delivering data to the outputs is contained in this block, for instance an error correction code, when present.

Then there is the part for the Test Modes. Together with the user mode, there is also a way, "hidden" to the customer, to access both the matrix and the circuitry to analyze the behavior of the device when operating.

Access to this kind of investigation is not explained in the manual, therefore only the chipmaker knows the enabling code.

There are mainly two reasons to "reserve" such operations: (1) many of these tests must be performed exactly knowing the physics of the cells, since all the voltages are applied to the matrix directly from the outside, and the operator must mimic the correct timing sequence, which is usually carried out by the circuitry; (2) the information that can be inferred can be quite sensitive, and the customer is not always able to analyze it under the right perspective. In Test Mode it is possible to directly access all the memory cells of the matrix, the references and the UPROM, to read, program and erase them using voltages provided by external generators.

The last function of the block under analysis refers to a particular synchronous read mode known as Burst Mode. The access time is related to the technological process used to build the device and to the complexity of the device itself. As we have seen, access time is in the range of 50÷100 ns. This value is constrained by the single asynchronous read. A way to speed up the reading phase is to access a higher number of bits at the same time and then to send outside a set of these bits while, in parallel, a new read is executed. In this way, after an initial latency time, fast reads can be achieved owing to the fact that more bits have been "prepared" at the beginning.

In this way, the gap between consecutive read operations can be as small as 15 ns. The limit in this case is given by the speed of the output buffer. Burst read necessarily requires a clock that times the operations; furthermore it cannot be random, but the addresses are scanned in sequential order.

The last block to consider is the I/O Pads. Task of the output buffers is to deliver to the external world the data acquired during a read operation. The output load is always constituted by a capacitance, whose value may range from 30 pF to 100 pF. The structure of the buffer can be roughly reduced to an inverter, whose pull-up charges the output if the cell is erased, or whose pull-down discharges the output in case of programmed cell. The design of the output buffer is complicated by the fact that such charge and discharge should occur as fast as possible: to achieve this result, a great amount of current is required and therefore very large output transistors, thus worsening area occupation issues. Furthermore it is important to control that the current used by the output buffers is not causing problems, either generated by the noise induced on the power supplies or caused by undesired coupling through the substrate with al the other circuits inside the device. In case of 16 simultaneous switching buffers, up to 1A or more of peak current might arise during the first nanoseconds.

3. MATRIX AND DECODERS

Let's take a closer look to the way the cells are connected to form the matrix. Figure 6-6 shows a typical arrangement for a NOR-type Flash process.

The represented portion of the sector contains, starting from the left, two columns of cells, one column of ground, 10 columns of cells, one column of ground and 2 columns of cells.

The structure of the entire sector is a repetition of what can be seen in Figure 6-6. There are 10 columns of cells between two columns of ground. The columns of cells constitute drain connections, while the task of the columns of ground is to collect the sources of the cells.

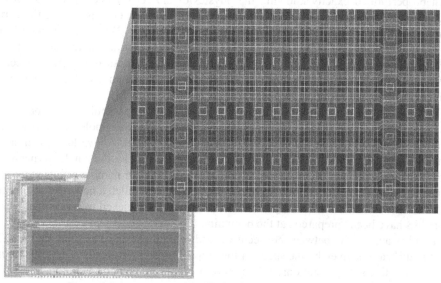

Figure 6-6. Arrangement of the Flash cells to form a NOR-type matrix.

Figure 6-7 shows the electrical scheme of the portion of the matrix previously described.

Besides the bit line or column, the source diffusion (Source Line) is shown, which is located to the opposite side with respect to drain contacts and which, through the metal of ground labeled Ground Line, is biased by the circuits located externally to the sector just for this task. The Word Line is the collection of the control gates of the cells, while the drain contacts connect the drain of the cells themselves to the bit line. The bits that constitute the byte or the word are taken at the same distance from the columns of ground, therefore, as shown for instance in the figure, the cells of the first column after the column of ground can all be bit 0 of different bytes, those of the second columns can all be bit 1 etc. In this way, activating the same row, the bits of the same byte are brought to the sense amplifier through the column decoding on a complex path (metal, contacts and active areas to cross) that is equal for all of them, thus reducing the parasitic effects to a common mode issue. Figure 6-8 shows the connection of the local row decoders to the cells of the sector. A zoom-in of a portion of the layout is depicted, indicating transistors M0, M1 and M2. In the last frame of the same figure, the electrical schematic of one switch used to pass the voltages of the global row to the local one (as already shown in Figure 6-4) is drawn to ease the comprehension of the represented layout portion.

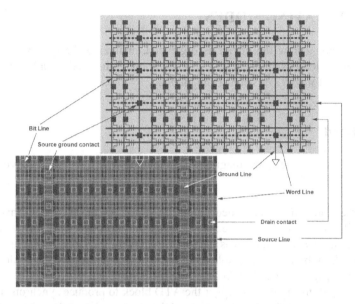

Figure 6-7. The electrical scheme corresponding to the portion of NOR sector.

Figure 6-8 shows also the way the local column decoding interfaces to the matrix; again, layout and related transistors schematic are shown.

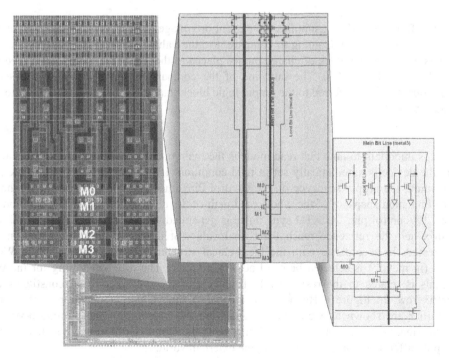

Figure 6-8. The position of the local row decoders in the matrix, the layout, the transistors and the overall electrical schematic.

4.   OPERATING MODES

After describing how the device is organized, analyzing the blocks required for its correct behavior, let's examine carefully the main operating mode of a Flash memory and the most used test mode. Let's start from the read.

4.1   Read

Figure 6-9 shows the block diagram of the device where the blocks involved in read mode are highlighted.

Of course read can be performed on all the sectors of the matrix; in this example, a memory cell inside one of these sectors is taken into account. In reality, read occurs in parallel on a number of cells that is defined by the way the device communicates to the external world: by byte, word, double word etc. The operation described below should be thought of as repeated in parallel on a certain number of cells.

As we learned, the addresses trigger the ATD block to produce synchronism signals as they propagate towards the decoders, both global and local, to select the sector, the row and the column. The band gap circuit provides the signals required to stabilize the read voltage to make it as independent as possible from supply voltage, temperature and process variations.

The Reference Matrix supplies the current or voltage to be used as a reference for the sense amplifier, which is going to compare it to the value read from the matrix cell. The row is biased by the block of charge pumps placed to the left of the global row decoders, which are called VPCX. The outcome of the comparison and decision of the sense amplifier is then transferred to the output logic block that in turn will pass the result to the output buffers.

It is important to note that read mode is the main operation of the memory: at power-up, the device is automatically set in read condition. If we need to bring it in a modify condition, the proper sequence on the Control Pins must be applied. Figure 6-10 shows the simulation snapshot of the whole read path of the device. Following the variation on the CE# control pin, the ATD sync pulse is generated. The Local Bit Lines of both the matrix and the reference reach their final value; at the same time, the word line is addressed. In accordance with the hierarchical structure described above, Main Word Line (in metal2) is selected before Local Word Line. The different timing for the two signals is related to the resistivity ($\sim 10 \, \Omega / \square$) of the polysilicon that constitutes the matrix row. In Figure 6-10 the input signals to the voltage comparator of the sense amplifier are shown as well: these signals are the outcome of the current-to-voltage conversion. Once the current of the cell is read, the Output Buffer Enable signal lets the output buffer communicate the data to the external world.

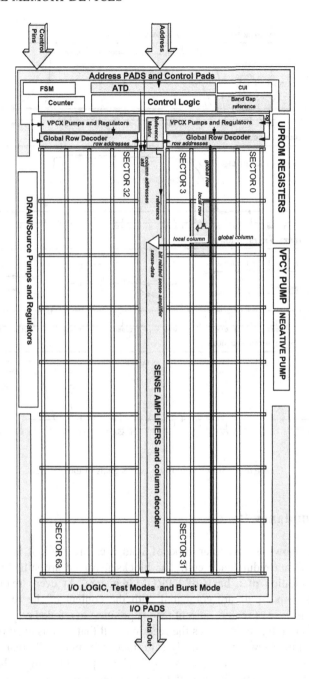

Figure 6-9. Blocks involved in the read path.

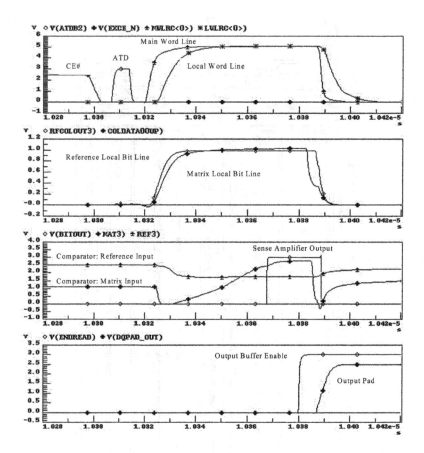

Figure 6-10. Simulation of the main signals of the read path.

4.2   Redundancy Read

Figure 6-11 shows a particular type of read, i.e. the read of a redundancy cell, specifically a column redundancy cell. The blocks involved are still those of a normal read, with the addition of a bank of UPROM registers. At every read, the addresses presented on the input pins are compared with the content of the registers inside the UPROM block. In case of match, current read is pointing to a failed cell.

The internal circuitry re-addresses the read to a cell that, in this example, belongs to a redundancy column. As we have already said, there are more cells than those normally addressable; these are used as a "spare wheel". The important thing is that the customer shouldn't notice this change. In fact, if the operations were sequential, read would be activated by the addresses at the same moment when the comparison of them with the content of the UPROM registers takes place. If a re-mapping of the address with a spare one were required, some time would be lost to turn off the current path and to restart with a new read process.

This would imply a downright distinction between "normal" and redundant devices because of the different access time on some memory locations. To overcome this issue, several circuital tricks have been implemented.

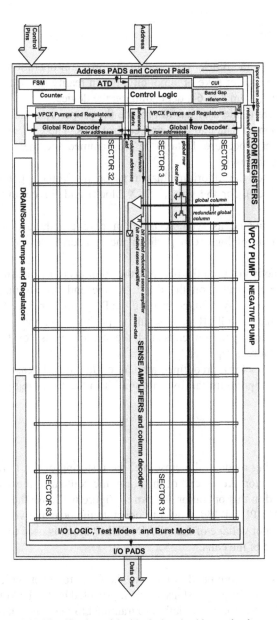

Figure 6-11. Identification of the blocks involved in a redundancy read.

The workaround shown in Figure 6-11 consists in having an additional sense amplifier dedicated to the redundancy column (R). Therefore at every new read operation, both the normal and the redundancy column are addressed, of course at the corresponding row. In this way the decision on the data to be output, through the multiplexer shown, can be made with more ease, since the time for it has been shifted forward in the event chain. If redundancy is activated, the data read from the redundancy sense is used; otherwise the normal data is output.

4.3   Program

Figure 6-12 shows the blocks involved in the program operation. Also in this case a single cell is considered for sake of simplicity, but in reality the operation takes place on more cells in parallel, from 8 to 64, depending on the device.

To activate program, the user has to provide the proper command to the device using the right combination of data / address and control pins. The CUI decodes the command activating the proper procedure. Data to be programmed and their related addresses are provided as well by the user.

The FSM executes the algorithm that turns on the voltage generator for the drain (DRAIN Pumps block), for the gate (VPCX block) and for column decoding (VPCY PUMP block).

UPROM are activated as well because, in case a redundancy location must be programmed, the control logic can perform the re-address towards the corresponding redundancy column.

The algorithm activates both the row and the columns to be biased and then it applies, for the defined time, a pulse on the drain. At the end of this pulse, gate voltage is modified to perform a verify read, comparing the Program Verify cell with the content of the programmed matrix cell. If read result is positive, i.e. the cell is considered as programmed, program operation ends and the device waits for the following input, otherwise the procedure is repeated.

The process iterates for a predefined amount of attempts. In case the limit is reached, the device signals the program fail by setting a bit in the status register that the user can read.

4.4   Erase

The third operating mode is the erase, i.e. the restoring of the initial condition for of all the cells in the sector. At the end of the erase, the content of these cells is a logic "1". The blocks involved in this operation, shown in Figure 6-13, are the same as in program, plus the NEGATIVE PUMP block used to generate the negative voltage to bias the gate of the cells. The output voltage of the DRAIN PUMP block is used to bias the body of the sector and the source of the cells.

Also in this case the command is input to the CUI, together with the address of the sector to be erased. The following steps are then executed by the FSM to perform erase.

All the cells of the sector are initially programmed to make the starting point uniform, considering that the erase pulses are applied to all the cells of the sector and otherwise these cells may have any logic value.

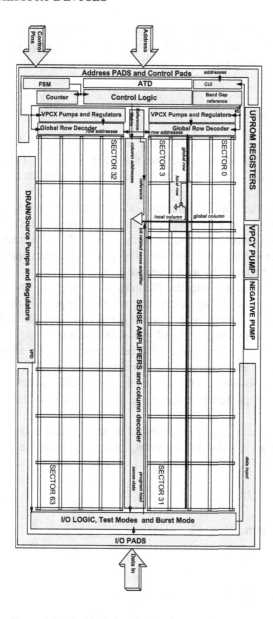

Figure 6-12. The blocks involved in the program operation.

This *preconditioning* operation is performed by the state machine that scans the rows and the columns of the sector and generates the required drain pulses. Then the erase starts: the cells are erased applying 10 ms pulses, and it usually takes some tens of them; after every pulse a verify on the whole sector is performed, scanning all the cells.

If even a single cell is verified as not erased, by comparing it with the Erase Verify cell, a new erase pulse is applied.

Then the recover of the depleted cells procedure, previously described, is performed, to recover both depleted and low threshold cells.

Both in this case and in program, described in the previous paragraph, the ATD block is not enabled. Since internal algorithms are executed during modify operations, an internal clock is activated to pace the various phases of the procedure.

5.   DMA TEST

For sake of completeness, let's show what is probably the most used test to characterize, analyze and study the cells of the matrix (Figure 6-14). DMA, acronym for Direct Memory Access, is the test mode that allows to directly access every single cell of the matrix.

All the cells are, of course, singularly addressable when read or programmed. In DMA, the sense amplifier is disconnected and a transistor is turned on to connect the column decoder, and therefore the drain of the cell, with the corresponding output pad. Applying the proper voltages in value and timing, the electrical characteristic of the cell can be drawn, perturbed only by the column decoding.

The row can be either activated by the normal decoding or driven by the outside by means of another test mode using an external generator to provide the voltage required by the analysis. This test mode can be used to get the distributions like those shown in Figure 6-5, which are fundamental to successfully manufacture a device.

The making of a device containing millions of memory cells, all apparently equal but indeed different, is a challenging task. Process variations that result in geometrical differences, different positions of the cells in the array, with respect to sector edges and to the ground lines, result in the end in variations in the behavior of the single cells.

The typical value of equivalent capacity of a memory cell is around 0.7 fF, where the charge is stored for a typical threshold voltage shift of 3 V and therefore it means that a programmed cell has got, on its floating gate, something like:

$$Q = 0.7\text{fF} \cdot 3\text{V} = 2.1\text{fC} \qquad (1)$$

That is, in terms of number of electrons

$$2.1\text{fC} \cong 2.1 \cdot 10^{-15} \cdot 1.6 \cdot 10^{19} = 33,600 \text{ electrons} \qquad (2)$$

Such a small number of electron does not help designing and manufacturing such devices. Hence a statistical FG cell model, as mentioned in the previous chapters, and a static analysis of memory, whose main tool is the DMA, are of highest interest

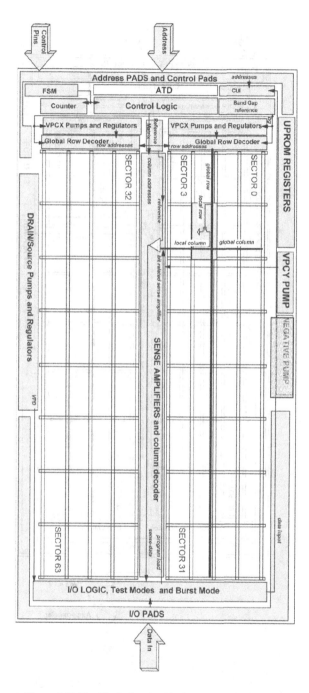

Figure 6-13. The blocks involved in the erase operation.

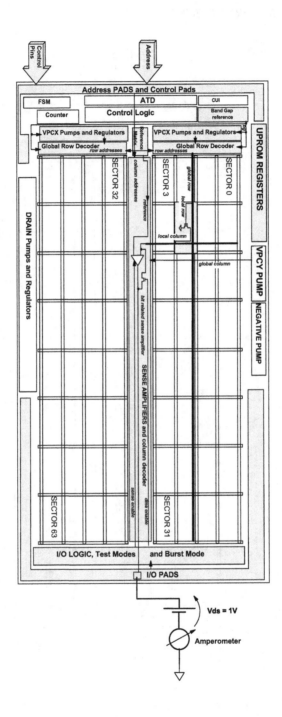

Figure 6-14. DMA test mode

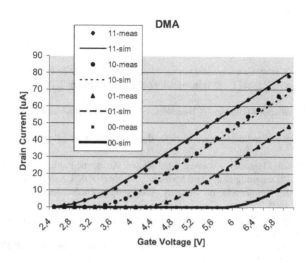

Figure 6-15. Current / voltage characteristics of memory cells measured in DMA.

Figure 6-15 shows the current/voltage characteristics measured in DMA of the matrix cells in case of multilevel device; in particular a memory able to store two bits per cell is considered (four distributions). In the same figure, four characteristics are shown together with the corresponding simulations, one for every distribution.

ACKNOWLEDGEMENT

The authors wish to thank Marcello Carrera and STMicroelectronics for providing the layout shown throughout the chapter. The authors also wish to thank Stefano Commodaro for reviewing this chapter.

Giovanni Campardo had been Flash Design Manager and he is now working in the System in Package field. He is author of more than 80 patents and publications among which: *FLASH MEMORIES* edited by P. Cappelletti et al., Kluwer Academic Publishers 1999, *PROGETTAZIONE DI MEMORIE NON VOLATILI*, Franco Angeli Editore, F. Angeli, collana scientifica, serie di Informatica. ISBN 88-464-2080-2. He was Co-Guest Editor for the Proceedings of the IEEE, April 2003, for the *Special Issue on Flash Memory Technology*.

Rino Micheloni is Product Development Manager of the multilevel products inside the Flash Storage Media Division of STMicroelectronics, Agrate Brianza, Italy. He is coauthor of the book *FLASH MEMORIES* edited by P. Cappelletti et al., Kluwer Academic Publishers, 1999. He was Co-Guest Editor for the Proceedings of the IEEE, April 2003, for the *Special Issue on Flash Memory Technology*.

REFERENCES

General

1. Lai S., "Flash memories: where we were and where we are going", *IEDM Tech Dig*, p. 971, 1998.
2. Pavan P., Bez R., Olivo P., Zanoni E., "Flash memory cells – An overview", Proc of the IEEE, 85, p.1248, 1997.
3. Betty Prince, "Semiconductor Memories. A Handbook of Design Manufacture and Application", Wiley & Sons, 1993.
4. B. Riccò et al., "Nonvolatile multilevel memories for digital applications", Proc. IEEE, vol. 86, pp. 2399-2421, Dec. 1998.
5. P. Cappelletti et al. "Flash Memories", Norwell, MA: Kluwer Academic Publishers, 1999.
6. B. Dipert and M. Levy, "Designing with Flash Memory", ANNABOOKS, San Diego 1994.
7. IEEE 1995 Nonvolatile Semiconductor Memory Workshop, "Flash Memory Tutorial", Monterey, California, August 14, 1995.
8. Campardo G., Micheloni R., "Special issue on flash memory technology", Proceedings of the IEEE, Vol. 91 (4), April 2003, pp. 483- 488.
9. Samuel Tuan Wang, "On the I-V characteristics of Floating-Gate Mos transistors", IEEE Transaction on electron devices, Vol ED-26, No 9, September 1979.
10. M. Ohkawa el al., "A $9.8mm^2$ Die size 3.3V 64Mb Flash memory with FN-NOR type Four level cell", IEEE, Journal of Solid-State Circuit, vol. 31, no. 11, Nov. 96, p. 1584.
11. P. L. Rolandi, et al.: "1M-cell 6b/cell analog flash memory for digital storage", IEEE ISSCC Dig. Tech. Papers, Feb. 1998, pp. 334-335.
12. R. Micheloni, O. Khouri, S. Gregori, A. Cabrini, G. Campardo, L. Fratin, G. Torelli, "A 0.13-µm CMOS NOR Flash Memory Experimental Chip for 4-b/cell Digital Storage", Proc. 28th European Solid-State Circuit Conference (ESSCIRC), Sept. 2002, pp. 131-134.
13. Conci A., Faldarini A., Fumagalli G., Girardi A., Pesare M., Tecli N., Zucchinali M., "Current criticalities and innovation perspectives in flash memory design Automation", Proceedings of the IEEE, Vol. 91 (4), April 2003, pp. 581- 593.
14. R. Gregorian, G. C. Temes, "Analog Mos integrated circuits for signal processing", J. Wiley & Sons, 1986, pp. 558-559.
15. Campardo G., Scotti M., Scommegna S., Pollara S., Silvagni A., "An overview of flash architectural developments", Proceedings of the IEEE, Vol. 91 (4), April 2003, pp. 523- 536.
16. G. Campardo and R. Micheloni "Architecture of non volatile memory with multi-bit cells" Elsevier Science, Microelectronic Engineering, Volume 59, Issue 1-4, November 2001, pp. 173-181.
17. M. Bauer et al., "A multilevel-cell 32Mb Flash memory", in 1995 IEEE Int. Solid-State Circuits Conf. Dig. Tech. Pap., Feb. 1995, pp. 132–133.

Process aspects

18. Ginami G., Canali D., Fattori D., Girardi G., Scintu P., Tarchini L., Tricarico D., "Survey on flash technology with specific attention to the critical process parameters related to manufacturing " Proceedings of the IEEE, Vol. 91 (4), April 2003, pp. 503- 522.
19. Bez R., Camerlenghi E., Modelli A., Visconti A., "Introduction to flash memory", Proceedings of the IEEE, Vol. 91 (4), April 2003, pp. 489- 502.
20. Stanley Wolf, "Silicon Processing For the VLSI ERA volume 2: Process integration", LATTICE PRESS, Sunset Beach, California, 1990.

Reliability

21. Dunn C., Kay C., Lewis T., Strauss T., Screck J., Hefley P., Middendorf M., San T., "Flash EEPROM disturb mechanism", Proc. Int. Rel. Phys. Symp., p.299, 1994.
22. Cappelletti P., Bez R., Cantarelli D., Fratin L., "Failure mechanisms of Flash cell in program/erase cycling", IEDM Tech. Dig., p.291, 1994.
23. Ielmini D., Spinelli A., Lacaita A., Modelli A., "Statistical Model of reliability and scaling projections for Flash memories", IEDM Tech. Dig., 2001.
24. A. Brand et al., "Novel Read Disturb Failure Mechanism Induced By FLASH Cycling", 1993, IEEE IRPS.

25. T.C.Ong et al., "ERRATIC ERASE IN ETOX^{TM} FLAH MEMORY ARRAY", VLSI Symposium on Technology, 7A-2, pagg. 83-84, 1993.
26. Cheng-Sheng Pan et al., "Physical Origin of Long-Term Charge Loss in Floating-Gate EPROM with an Interpoly Oxide-Nitride-Oxide Stacked Dielectric", IEEE Electron Device Letters, Vol. 12, No. 2, February 1991.
27. Y. Igura et al., "New Device Degradation Due to "Cold Carriers Created by Band-to Band Tunneling", IEEE Electro Device Letters, VOL. 10, NO. 5, MAY 1989.
28. Verma et al., "Reliability Performance of ETOX based FLASH memories", 1988, IEEE/IRPS.
29. Chimenton A., Pellati P., Olivo P., "Overerase phenomena: an insight into flash memory reliability", Proceedings of the IEEE, Vol. 91 (4), April 2003, pp. 617- 626.

High Voltage circuitry

30. M. Maccarrone et al., "Program load adaptative voltage generator for Flash memories", Journal of Solid State Circuit, Vol. 32, No 1, Jan 97, p. 100.
31. O. Khouri, R. Micheloni, and G. Torelli, "Very fast recovery word-line voltage regulator for multilevel non-volatile memories", in *Proc.* Third IMACS/IEEE Int. Multiconference Circuits, Communications and Computers, Athens, Greece, June 1999, pp. 3781–3784.
32. O. Khouri, R. Micheloni, A. Sacco, G. Campardo, and G. Torelli, "Program word-line voltage generator for multilevel Flash memories", in Proc. 7th IEEE Int. Conf. on Electronics, Circuits, and Systems, Dec. 2000, vol II, pp. 1030–1033.
33. Motta I., Ragone G., Khouri O., Torelli G., Micheloni R., "High-voltage management in single-supply CHE nor-type flash memories", Proceedings of the IEEE, Vol. 91 (4), April 2003, pp. 554- 568.
34. Grossi M., Lanzoni M., Ricco B., "Program Schemes For Multilevel Flash Memories",, Proceedings of the IEEE, Vol. 91 (4), April 2003, pp. 594- 601.

Erase

35. Frohman-Bentchkowsky D., "Memory behavior in a floating-gate avalanche-injection MOS (FAMOS) structure", Applied Physics Letters, 18, p.332, 1971.
36. Frohman-Bentchkowsky D., "FAMOS – a new semiconductor charge storage device", Solid State Electronics, 17, p.517, 1974.
37. Yoshikawa K., Yamada S., Miyamoto J., Suzuki T., Oshikiri M., Obi E., Hiura Y., Yamada K., Ohshima Y., Atsumi S., "Comparison of current Flash EEPROM erasing methods: stability and how to control", IEDM Tech. Dig., p.595, 1992.
38. S. Yamada, "A Self-Convergence Erasing Scheme for a Simple Stacked Gate FLASH EEPROM", 1991, IEEE IEDM, 11.4.1.
39. A. Montalvo, "Flash EEPROM array with paged erase architecture", USA Patent N 5,126,808.
40. S. Atsumi et al., "A channel-erasing 1.8V-Only 32-Mb NOR flash EEPROM with a bitline direct sensing scheme", IEEE J. Solid-State Circuits, vol. SC-35, pp. 1648–1654, Nov. 2000.
41. T. Jnbo et al., "A 5V only 16Mb Flash memory with sector Erase Mode", IEEE Journal of Solid State Circuits, Vol 27, N 11, November 1992.
42. C. Auricchio et al., "A triple-weel architecture for low voltage operation in submicron CMOS devices", (Eds.), Proc. ESSDERC 96, Bologna, Italy, p. 613, 1996.
43. V. N. Kynett et al., "A 90-ns one-million erase/program cycle 1-Mbit Flash memory", IEEE J. Solid-State Circuits, vol. SC-24, pp. 1259-1264, Oct. 1989.

Hierarchical Decoders

44. G. Campardo, et al., "40-mm^2 3-V-only 50-MHz 64-Mb 2-b/cell CHE NOR Flash memory", IEEE J. Solid-State Circuits, vol. SC-35, no. 11, Nov. 2000, pp.1655-1667.
45. A. Silvagni et al., "Modular Architecture For a Family of Multilevel 256/192/128/64MBIT 2-Bit/Cell 3V Only NOR FLASH Memory Devices", ICECS 2001 The 8th IEEE international Conference on Electronics, Circuits and System, September 2-6, 2001- Malta.
46. H. Arakawa, "Address Decoder Circuit for Non-Volatile Memory", United States Patent 5,039,882, Aug 13, 1991.
47. A. Umezawa et al., "A 5V-Only operation 0.6 μm Flash EEPROM with Row decoder scheme in triple-well structure", IEEE Journal of Solid State Circuits, Vol. 27, N 11, November 1992.

48. Y. Miyawaki et al., "A New Erasing and Row decoding scheme for low supply voltage operation 16-Mb/64-Mb Flash Memories", IEEE Journal of Solid State Circuits, Vol. 27, N 4, April 1992.

49. S. Atsumi et al., "16Mb flash memory with row-decoding scheme", 1996 IEEE Int. Solid-State Circuits Conf. Dig. Tech. Papers, Feb. 1996, pp. 42–43.

50. R. Micheloni et al., "Hierarchical Sector Biasing Organization for Flash Memories", 2000 IEEE International Workshop on Memory Technology, Design and Testing (MTDT 2000), San Jose, California (USA), August 2000, pp. 29-33.

Sense amplifier

51. A. Pierin, S. Gregori, O. Khouri, R. Micheloni, G. Torelli, "High-Speed Low-Power Sense Comparator for Multilevel Flash Memories" in Proc. 7th Int. Conf. Electronics, Circuits and Systems, vol. II, Dec. 2000, pp. 759 – 762.

52. S. H. Dhong et al., "High Speed Sensing Scheme for CMOS DRAM's", IEEE Journal of Solid-State Circuits, vol. 23, no. 1, February 1988, pp. 34-40.

53. T. Kawahara et al., "Bit-Line camped sensing multiplex and accurate high voltage generator for 0.25 μm flash memories", in 1996 IEEE Int. Solid-State Circuits Conf. Dig. Tech. Pap, pp. 38-39.

54. Y. Konishi et al., "Analysis of Coupling Noise Between Adjacent Bit Lines in Megabit DRAM's", IEEE Journal of Solid-State Circuits, vol. 24, no. 1, February 1989.

55. G. Campardo, "Circuit for sensing the state of matrix cell in NMOS EPROM memories including an offset current generator", USA patent No. 4,949,307.

56. Micheloni R., Crippa L., Sangalli M., Campardo G., "The flash memory read path: building blocks and critical aspects", Proceedings of the IEEE, Vol. 91 (4), April 2003, pp. 537- 553.

Charge Pumps

57. O. Khouri, S. Gregori, R. Micheloni, D. Soltesz, and G. Torelli, "Low output resistance charge pump for Flash memory programming", 2001 IEEE Proc. Int. Workshop on Memory Technology, Design and Testing, San Jose, CA (USA), Aug. 2001, pp. 99–104.

58. J. T. Wu and K.L. Chang, "MOS charge pumps for low-voltage operation", IEEE J. Solid-State Circuits, vol. 33, Apr. 1998, pp.592-597.

59. T. Tanzawa and T. Tanaka, "A dynamic analysis of the Dickson charge pump circuit", IEEE J. Solid-State Circuits, vol. SC-32, no. 8, Aug. 1997, pp.1231-1240.

60. J. Dickson, "On-chip high voltage generation in MNOS integrated circuits using an improved voltage multiplier technique", IEEE J. Solid-State Circuits, vol. SC-11, no. 3, June 1976, pp.374-378.

61. G. Di Cataldo and G.Palumbo, "Double and triple charge pump for power IC dynamic models which take parasitic effects into account", IEEE Trans. Circuits Syst., vol. CAS-40, pp. 92-101, Feb. 1993.

62. C. -C Wang and J. Wu, "Efficiency improvement in charge pump circuits", IEEE J. Solid-State Circuits, vol. SC-32, June 1997, pp. 852-860.

63. P. Favrat, P. Deval, and M. J. Declercq, "A high-efficiency CMOS voltage doubler", IEEE J. Solid-State Circuits, vol. SC-33, pp. 410-416, March 1998.

64. N. Otsuka and M. A. Horowitz, "Circuit Techniques for 1.5-V Power Supply Flash Memory", IEEE Journal of Solid-State Circuits, vol. 32, no. 8, August 1997.

65. H. Morimura and N. Shibata, "A Step-Down Boosted-Wordline Scheme for 1-V Battery-Operated Fast SRAM's", IEEE Journal of Solid-State Circuits, vol. 33, no. 8, August 1998.

66. T. Tanzawa and S. Atsumi, "Optimization of word-line booster circuits for low voltage flash memories", IEEE J. Solid-State Circuits, vol. SC-33, pp. 410-416, March 1998.

Standby

67. Kobayashi el al., "A 3.3V-Only 16 MB DINOR Flash Memory", ISSCC95, FLASH MEMORY SESSION.

68. B. Venkatesh et al., "A 55ns 0.35 μm 5V-Only 16M Flash Memory with Deep-Power-Down", ISSCC'96.

69. R. Micheloni, I. Motta, O. Khouri, and G. Torelli: "Stand-by low-power architecture in a 3-V only 2-bit/cell 64-Mbit Flash memory", in Proc. 8th IEEE Int. Conf. Electronics, Circuits, and Systems, Sept. 2001, vol. II, pp. 929-932.

Logic

70. C. Clare, "Designing logic system using state machines", Mc Graw Hill, 1973.
71. T. Nakayama, "A 60ns 16Mb Flash EEPROM with Program and Erase Sequence Controller", ISSCC91, paper FA 16.1, pp. 260-261.
72. Silvagni A., Fusillo G., Ravasio R., Picca M., Zanardi S., "An overview of logic architectures inside flash memory devices", Proceedings of the IEEE, Vol. 91 (4), April 2003, pp. 569- 580.

Error Correction Code

73. R. Vancu et al., "A 35ns 256k CMOS EEPROM with error correcting circuitry," in *1990* IEEE Int. Solid-State Circuits Conf. Dig. Tech. Pap., Feb. 1990, pp. 64-65.
74. T. Tanzawa et al., "A compact on-chip ECC for low cost Flash memories," IEEE J. Solid-State Circuits, vol. 32, no. 5, pp. 662-669, May 1997.
75. S. Gregori, O. Khouri, R. Micheloni, and G. Torelli, "An error control code scheme for multilevel Flash memories," in Records 2001 IEEE International Workshop on Memory Technology, Design and Testing, Aug. 2001, pp. 45-49.
76. Gregori et al., "On-chip error correcting techniques for new-generation flash memories", Proceedings of the IEEE, Vol. 91 (4), April 2003, pp. 602- 616.

Input/Output Circuitry

77. H. B. Bakoglu, "Circuits interconnections and packaging for VLSI", Addison Wesley Pub. 1990.
78. K. Ming-Dou, W. Chung-Yu, W. Tain-Shun, "Area-Efficient Layout Design for CMOS Output Transistors", IEEE, Trans. On Electron Devices, vol. 44, no. 4, April 1997.

Acknowledgments

The authors would like to thank many great people in STMicroelectronics for their comments and fruitful discussions: L. Albani, R. Annunziata, R. Bez, E. Camerlenghi, D. Cantarelli, P. Cappelletti, M. Cuozzo, G. Ghidini, T. Ghilardi, P. Pansana, F. Pellizzer, P.Zuliani, from the Agrate plant, A. Conte, G. Crisenza, S. Lo Giudice, and M. Gaibotti from Catania plant. C. Maiorca from Mentor Graphics helped in the engineering effort. The authors would also like to acknowledge the efforts of the students from the University of Modena and Reggio Emilia who enthusiastically worked for this project: F. Gattel (now with STMicroelectronics), R. Compagni (now with F.lli Zanasi) and S. Bertulu (now with M. D. Microdetectors), E. Gambuzzi, M. Terzi. Finally the authors acknowledge the valuable contribution of G. Torelli, Professor at the University of Pavia. The constant support of Prof. C. Canali (University of Modena and Reggio Emilia) and of Prof. E. Zanoni and Prof. A. Paccagnella (University of Padova) has been greatly appreciated. Paolo Pavan especially thanks Dr. Boaz Eitan and many other friends in Saifun Semiconductors, Israel.